Advances in Intelligent Systems and Computing

Volume 396

Series editor

Janusz Kacprzyk, Polish Academy of Sciences, Warsaw, Poland
e-mail: kacprzyk@ibspan.waw.pl

About this Series

The series "Advances in Intelligent Systems and Computing" contains publications on theory, applications, and design methods of Intelligent Systems and Intelligent Computing. Virtually all disciplines such as engineering, natural sciences, computer and information science, ICT, economics, business, e-commerce, environment, healthcare, life science are covered. The list of topics spans all the areas of modern intelligent systems and computing.

The publications within "Advances in Intelligent Systems and Computing" are primarily textbooks and proceedings of important conferences, symposia and congresses. They cover significant recent developments in the field, both of a foundational and applicable character. An important characteristic feature of the series is the short publication time and world-wide distribution. This permits a rapid and broad dissemination of research results.

Advisory Board

More information about this series at http://www.springer.com/series/11156

Rituparna Chaki · Agostino Cortesi
Khalid Saeed · Nabendu Chaki
Editors

Advanced Computing and Systems for Security

Volume 2

 Springer

Editors
Rituparna Chaki
University of Calcutta
Kolkata, West Bengal
India

Agostino Cortesi
Università Ca' Foscari
Venice
Italy

Khalid Saeed
Faculty of Computer Science
Bialystok University of Technology
Białystok
Poland

Nabendu Chaki
University of Calcutta
Kolkata, West Bengal
India

ISSN 2194-5357   ISSN 2194-5365   (electronic)
Advances in Intelligent Systems and Computing
ISBN 978-81-322-2651-2   ISBN 978-81-322-2653-6   (eBook)
DOI 10.1007/978-81-322-2653-6

Library of Congress Control Number: 2015951344

Springer New Delhi Heidelberg New York Dordrecht London

Springer (India) Pvt. Ltd. is part of Springer Science+Business Media (www.springer.com)

Preface

The Second International Doctoral Symposium on Applied Computation and Security Systems (ACSS 2015) took place during May 23–25, 2015 in Kolkata, India. The University of Calcutta collaborated with Ca' Foscari University of Venice, Bialystok University of Technology, and AGH University of Science and Technology, Poland, to make ACSS 2015 a grand success.

The symposium aimed to motivate Ph.D. students to present and discuss their research works to produce innovative outcomes. ACSS 2015 invited researchers working in the domains of Computer Vision & Signal Processing, Biometrics-based Authentication, Machine Intelligence, Algorithms, Natural Language Processing, Security, Remote Healthcare, Distributed Systems, Embedded Systems, Software Engineering, Cloud Computing & Service Science, Big Data, and Data Mining to interact.

By this year, the post-conference book series are indexed by ISI Compendex. The sincere effort of the program committee members coupled with ISI indexing has drawn a large number of high-quality submissions from scholars all over India and abroad. A thorough double-blind review process was carried out by the PC members and by external reviewers. While reviewing the papers, reviewers mainly looked at the novelty of the contributions, at the technical content, at the organization, and at the clarity of presentation. The entire process of paper submission, review, and acceptance process was done electronically. Due to the sincere efforts of the Technical Program Committee and the Organizing Committee members, the symposium resulted in a suite of strong technical paper presentations followed by effective discussions and suggestions for improvement of each researcher.

The Technical Program Committee for the symposium selected only 37 papers for publication out of 92 submissions. During each session, the authors of each presented paper were given a list of constructive suggestions in a bid to improve upon their work. Each author had to incorporate the changes in the final version of the paper as suggested by the reviewers and the respective session chairs. The symposium Proceedings are organized as a collection of papers, on a session-wise basis.

We take this opportunity to thank all the members of the Technical Program Committee and the external reviewers for their excellent and time-bound review works. We thank all the sponsors who have come forward toward the organization of this symposium. These include Tata Consultancy Services (TCS), Springer India, ACM India, M/s Business Brio, and M/s Enixs. We appreciate the initiative and support from Mr. Aninda Bose and his colleagues in Springer for their strong support toward publishing this post-symposium book in the series "Advances in Intelligent Systems and Computing." Last but not least, we thank all the authors without whom the symposium would not have reached this standard.

On behalf of the editorial team of ACSS 2015, we sincerely hope that this book will be beneficial to all its readers and motivate them toward further research.

Rituparna Chaki
Agostino Cortesi
Khalid Saeed
Nabendu Chaki

Contents

About the Editors

Rituparna Chaki has been an Associate Professor in the A.K. Choudhury School of Information Technology, University of Calcutta, India since June 2013. She joined academia as faculty member in the West Bengal University of Technology in 2005. Before that she has served under Government of India in maintaining industrial production database. Rituparna has received her Ph.D. from Jadavpur University in 2002. She has been associated with organizing many conferences in India and abroad by serving as Program Chair, OC Chair or as member of Technical Program Committee. She has published more than 60 research papers in reputed journals and peer-reviewed conference proceedings. Her research interest is primarily in ad hoc networking and its security. She is a professional member of IEEE and ACM.

Agostino Cortesi received his Ph.D. degree in Applied Mathematics and Informatics from University of Padova, Italy, in 1992. After completing his post-doc at Brown University, in the US, he joined the Ca' Foscari University of Venice. In 2002, he was promoted to full professor of Computer Science. In recent past, he served as Dean of the Computer Science program, as Department Chair, and as Vice-Rector of Ca' Foscari University for quality assessment and institutional affairs. His main research interests concern programming languages theory, software engineering, and static analysis techniques, with particular emphasis on security applications. He has published over 100 papers in high-level international journals and proceedings of international conferences. His h-index is 15 according to Scopus, and 23 according to Google Scholar. Agostino served several times as a member (or chair) of program committees of international conferences (e.g., SAS, VMCAI, CSF, CISIM, ACM SAC) and he is in the editorial boards of the journals such as "Computer Languages, Systems and Structures" and "Journal of Universal Computer Science."

Khalid Saeed received the B.Sc. degree in Electrical and Electronics Engineering from Baghdad University in 1976, the M.Sc. and Ph.D. degrees from Wrocław University of Technology, in Poland in 1978 and 1981, respectively. He received his D.Sc. Degree (Habilitation) in Computer Science from Polish Academy of

Sciences in Warsaw in 2007. He is a Professor of Computer Science with AGH University of Science and Technology in Poland. He has published more than 200 publications—edited 23 books, journals and conference proceedings, eight text and reference books. He has supervised more than 110 M.Sc. and 12 Ph.D. theses. His areas of interest are biometrics, image analysis and processing, and computer information systems. He gave 39 invited lectures and keynotes in different universities in Europe, China, India, South Korea, and Japan. The talks were on biometric image processing and analysis. He received about 18 academic awards. Khalid Saeed is a member of more than 15 editorial boards of international journals and conferences. He is an IEEE Senior Member and has been selected as IEEE Distinguished Speaker for 2011–2016. Khalid Saeed is the Editor in Chief of International Journal of Biometrics with Inderscience Publishers.

Nabendu Chaki is a Senior Member of IEEE and Professor in the Department of Computer Science and Engineering, University of Calcutta, India. Besides editing several volumes in Springer in LNCS and other series, Nabendu has authored three textbooks with reputed publishers like Taylor & Francis (CRC Press), Pearson Education, etc. Dr. Chaki has published more than 120 refereed research papers in Journals and International conferences. His areas of research interests include image processing, distributed systems, and network security. Dr. Chaki has also served as a Research Assistant Professor in the Ph.D. program in Software Engineering at the Naval Postgraduate School, Monterey, CA, USA. He is a visiting faculty member in many universities including the University of Ca' Foscari, Venice, Italy. Dr. Chaki has contributed in SWEBOK v3 of the IEEE Computer Society as a Knowledge Area Editor for Mathematical Foundations. Besides being in the editorial board of Springer and many international journals, he has also served in the committees of more than 50 international conferences. He has been the founding Chapter Chair for ACM Professional Chapter in Kolkata, India since January 2014.

Part I
Signal Processing

Design and Development of Marathi Speech Interface System

Santosh Gaikwad, Bharti Gawali and Suresh Mehrotra

Abstract Speech is the most prominent and natural form of communication between humans. It has potential of being an important mode of interaction with computer. Man–machine interface has always been proven to be a challenging area in natural language processing and in speech recognition research. There are growing interests in developing machines that can accept speech as input. Normal person generally communicate with the computer through a mouse or keyboard. It requires training and hard work as well as knowledge about computer, which is a limitation at certain levels. Marathi is used as official language at government of Maharashtra. There is a need for developing systems that enable human–machine interaction in Indian regional languages. The objective of this research is to design and development of the Marathi speech Activated Talking Calculator (MSAC) as an interface system. The MSAC is speaker-dependent speech recognition system that is used to perform basic mathematical operation. It can recognize isolated spoken digit from 0 to 50 and basic operation like addition, subtraction, multiplication, start, stop, equal, and exit. Database is an essential requirement to design the speech recognition system. To reach up to the objectives set, a database having 22,320 sizes of vocabularies is developed. The MSAC system trained and tested using the Mel Frequency Cepstral Coefficients (MFCC), Linear Discriminative Analysis (LDA), Principal Component Analysis (PCA), Linear Predictive Codding (LPC), and Rasta-PLP individually. Training and testing of MSAC system are done with individually Mel Frequency Linear Discriminative Analysis (MFLDA), Mel Frequency Principal Component Analysis (MFPCA), Mel Frequency Discrete

S. Gaikwad (✉) · B. Gawali · S. Mehrotra
System Communication Machine Learning Research Laboratory (SCM-RL),
Department of Computer Science and Information Technology,
Dr. Babasaheb Ambedkar Marathwada University, Aurangabad,
Maharashtra, India
e-mail: santosh.gaikwadcsit@gmail.com

B. Gawali
e-mail: bharti_rokade@yahoo.co.in

S. Mehrotra
e-mail: mehrotra_suresh15j@gmail.com

© Springer India 2016   3
R. Chaki et al. (eds.), *Advanced Computing and Systems for Security*,
Advances in Intelligent Systems and Computing 396,
DOI 10.1007/978-81-322-2653-6_1

Wavelet Transformation (MFDWT), and Mel Frequency Linear Discrete Wavelet Transformation (MFLDWT) fusion feature extraction techniques. This experiment is proposed and tested the Wavelet Decomposed Cepstral Coefficient (WDCC) with 18, 36, and 54 coefficients approach. The performance of MSAC system is calculated on the basis of accuracy and real-time factor (RTF). From the experimental results, it is observed that the MFCC with 39 coefficients achieved higher accuracy than 13 and 26 variations. The MFLDWT is proven higher accuracy than MFLDA, MFPCA, MFDWT, and Mel Frequency Principal Discrete Wavelet Transformation (MFPDWT). From this research, we recommended that WDCC is robust and dynamic techniques than MFCC, LDA, PCA, and LPC. MSAC interface application is directly beneficial for society people for their day to day activity.

Keywords Human-computer interaction · MSAC · MFCC · PCA · WDCC · LPC

1 Introduction

Speech is the most natural and efficient form of exchanging information between human. Automatic speech recognition (ASR) is defined as the process of converting a speech signal to a sequence of words by means of an algorithm implemented by a computer program. Speech recognition systems help users who cannot be able to use the traditional Input and Output (I/O) devices [1, 2]. The man–machine interface using speech recognition has helpful ways to for enable the visually impaired and computer laymen to use the updated technologies [3]. There are growing interests in developing machines that can accept speech as input. Given the substantial research efforts in the speech recognition worldwide and the rate at which computer becomes faster and smaller, we can expect more applications of speech recognition. The concept of machines being able to interact with people in natural form is very interesting. It is desirable to have machine interaction in voice mode in one's native language. This is exclusively important in multilingual country such as India, where a majority of the people is not comfortable with speaking, reading, and listening English language [4]. Research in ASR by machine has attracted a great deal of attention over the past six decades. For centuries, researcher has tried to develop machines that can produce and understand speech as humans do so naturally in native language. Some successful applications of speech recognition are virtual reality, multimedia searches, auto-attendants, IVRS, natural language understanding and many more applications [5–7].

This paper is organized as follows: Sect. 2 describes the related work of this experiment. Section 3 describes the speech recognition with feature extraction techniques. Section 4 explains the design of the Marathi speech interface system. Section 5 describes the experimental results and their extensive application for talking calculator and Sect. 6 gives the concluding remark of paper followed by references.

2 Related Work

The main goal of speech recognition is to develop techniques and recognition sys-tems for speech input to the machine. Human beings are comfortable speaking directly with computers rather than depending on primitive interfaces such as key-boards and pointing devices. The primitive interfaces like keyboard and pointing devices require a certain amount of skill for effective usage. Use of mouse requires good hand-eye coordination. It is very difficult for visually handicapped person to use the computer. Moreover, the current computer interface assumes a certain level of literacy from the user. It expects the user to have certain level of proficiency in English apart from typing skill. Speech interface helps to resolve these issues [8]. Computers which can recognize speech in native languages enable the common man to make use of those benefits in education technology [9]. The researcher turns toward performance improvement of speech recognition system for significant innovation of real-time applications [10]. Many factors are affecting the speech recognition such as regional, sociolinguistic, or related to the environment. These create a wide range of variations that may not be modeled correctly (speaker, gender, speaking rate, vocal effort, regional accent, speaking style, nonstationary, etc.), especially when resources for system training are infrequent. Performance is affected by speech variability [11]. The majority of technological changes have been directed toward the purpose of increasing robustness of the recognition system [12].

N.S. Nehe et al., proposed an efficient feature extraction method for speech recogni-tion. The features were obtained Linear Predictive Codding (LPC) and Discrete Wavelet Transformation (DWT). The proposed approach provides effective (better recognition rate), efficient (reduced feature vector dimension) features. Continuous Density Hidden Markov Model (CDHMM) has been implemented for system classification. The pro-posed algorithms were evaluated using an isolated Marathi digits database in the presence of white Gaussian noise [13]. Rajkumar S. Bhosale et al., worked on speech-independent recognition system using multi-class support vector machine and LPC [14].

The research work for Indian languages in speech recognition has not yet grasped to a critical level for a real-time communication as compare to other languages of developed countries. Countable attempts to develop a speech recog-nition system had been attempted by HP Labs India and IBM research lab, Google, IIT Powai, CDAC Pune [15, 16].

However, there is lots of opportunity to develop a speech recognition system for Indian languages. To achieve such aspiring motivation, the research is to develop Marathi speech interface system for talking calculator application.

3 Speech Recognition and Feature Extraction Techniques

For the development of Marathi speech activated calculator (MSAC) system, the recognition of speech is necessary. The speech is recognized and then the action is performed. For the speech recognition system speech recording (database creation),

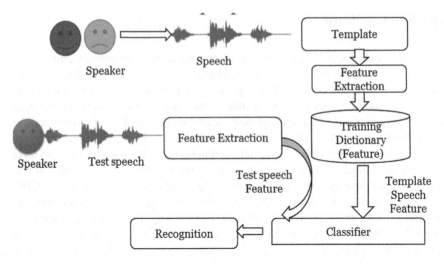

Fig. 1 Working of speech recognition system

feature extraction, training, classification, and testing are fundamental steps. The step by step flow diagram of speech recognition system is described in Fig. 1.

3.1 Feature Extraction Techniques

The MSAC system is trained and tested using three approaches of feature extraction, such as basic feature extraction techniques, fusion approach of feature extraction techniques and Wavelet Decomposed Cepstral Coefficients (WDCC): Proposed Approach. In the basic feature extraction techniques, the Mel Frequency Cepstral Coefficient (MFCC), Linear Discriminative Analysis (LDA), Principal Component Analysis (PCA), LPC, Rasta-PLP Analysis, and Discrete Wavelet Transformation were implemented.

(a) *Mel Frequency Cepstral Coefficient*

The enriched literature available on speech recognition, hence reported that the MFCC is most popular and robust technique for feature extraction [17, 18]. The MFCC is based on the known variation of the human ear's critical bandwidth frequencies with filters spaced linearly at low frequencies [19]. In this experiment, we extracted the 13 and 39 features of MFCC. Figure 2 shows the graphical representation of shows 39 Mel Frequency Cepstral coefficients (MFCC) for गणकयंत्र of first seven frames. The MFCC extracted the basic feature, variation of energy feature.

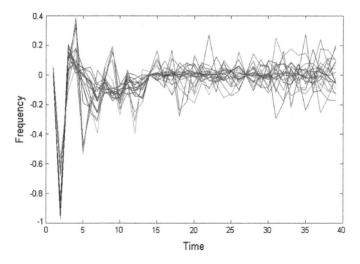

Fig. 2 Graphical representation of MFCC 39 coefficients of word गणकयंत्र

(b) *Linear Discriminative Analysis (LDA)*

LDA algorithm provides better classification compared to principal components analysis [20]. From the literature the Linear Discriminant Analysis (LDA) is commonly used technique for data classification and dimensionality reduction, but in this research we used for feature extraction. Figure 3 shows the graphical representation of extracted LDA feature of speech signal गणकयंत्र for first 10 frames.

The LDA feature is the combination of the Projection matrix feature, eigenvalues, Mean square representation error, Bias feature, and mean of training data.

(c) *Principal Component Analysis (PCA)*

The principal component analysis is the techniques for classification, but here we are used as feature extraction and dimension reduction. Figure 4 represents the graphical representation of interclass identification of PCA feature.

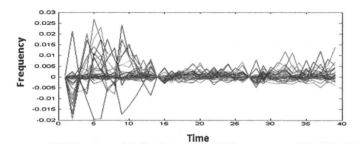

Fig. 3 The LDA feature set for the speech signal

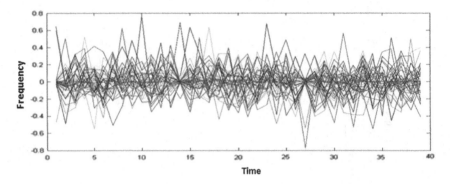

Fig. 4 PCA with class classification

The principal component analysis extracts the 30 feature which is the combinations of projection feature, in class variation, the mean and eigenvalues of the speech signal.

(d) *LPC*

The LPC is one of the robust and dynamic speech analysis techniques. In this research, we used the LPC for feature extraction. The graphical representation of the mean extracted LPC coefficient of speech signal गणकयंत्र is described in Fig. 5. The LPC 7 coefficients contain the Pitch, gain, and duration coefficient parameters of energy.

(e) *Rasta—PLP*

Rasta—PLP techniques is used for feature extraction. This extracted feature is the combination of the graphical band of voice signal. Figure 6 represents the graphical representation of the extracted Rasta PLP coefficients.

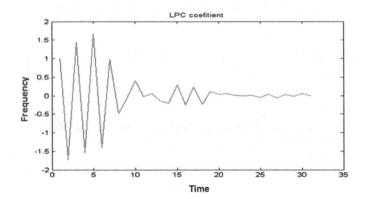

Fig. 5 Mean of extracted LPC coefficient of the speech signal

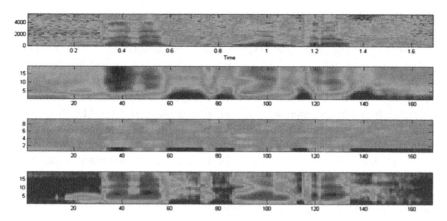

Fig. 6 RASTA-PLP coefficient of speech signal गणकयंत्र

The Rasta feature contains the pitch, gain and duration of energy and short-term noise coefficients.

(f) *DWT*

The wavelet series is just a sampled version of continuous wavelet transformation and its computation may consume a significant amount of time and resources, depending on the resolution required [21]. The DWT is also used for feature extraction as well as dimension reduction approach. The graphical representation of DWT approximation coefficient is described in Fig. 7.

DWT coefficient is calculated at approximation and detail level. Extracted DWT coefficients include the frequency variation of each frequency band in approximation level and energy variation with time duration in detail coefficients.

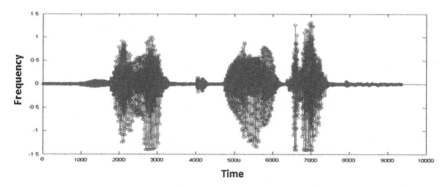

Fig. 7 The extracted approximation coefficient of the speech signal गणकयंत्र

3.1.1 Fusion-Based Feature Extraction Techniques

The fusion approach means combination of different techniques. Total 13 MFCC features were extracted and feature vector was formed. The formed feature vector was passed to fusion technique as an input. The detail fusion approach of different techniques with MFCC and their properties is explained in Table 1.

For the fusion-based approach, this research implemented the Mel Frequency Linear Discriminative Analysis (MFLDA), Mel Frequency Principal Component Analysis (MFPCA), Mel Frequency Discrete Wavelet Transformation (MFDWT), Mel Frequency Principal Discrete Wavelet Transformation (MFPDWT), and Mel Frequency Linear Discrete Wavelet Transformation (MFLDWT) fusion approach as feature extraction techniques. Figure 8 represents the graphical representation of extracted MFLDA feature for word गणकयंत्र"

3.1.2 WDCC: Proposed Approach

In proposed WDCC, the original speech signal is decomposed to second level. The packet coefficient offers different time, frequency representation qualities and consequently potential, for adaptation of the time series phenomenon. This strategy of decomposition offers richest analysis of signal [22–26]. In the WDCC techniques, the original speech signal is decomposed second level. The approximation and detail coefficient is a distinguished output from decomposition step. The DCT operation is performed on horizontal coefficient, which is fused with basic acoustic

Table 1 Fusion approach with MFCC and their properties

Sr. no	Name of fusion technique	Combination of techniques	Input feature vector	Output feature vector	Properties
1	MFLDA	Fusion of MFCC and LDA	13	02	It is used for dimension reduction
2	MFPCA	Fusion of MFCC and PCA	13	02	It is used for dimension reduction
3	MFDWT	Fusion of MFCC and DWT	13	01	The speed is fast as compare to other techniques
4	MFPDWT	Fusion of MFCC, PCA and DWT	13	01	It is used to reduce time complexity
5	MFLDWT	Fusion of MFCC, LDA and DWT	13	01	It is also used to reduce time complexity

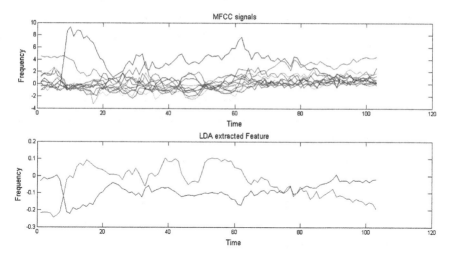

Fig. 8 The fusion approach of MFLDA (MFCC and LDA) of speech signal "गणकयंत्र"

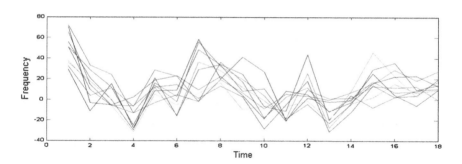

Fig. 9 The WDCC extracted 18 features of word "गणकयंत्र"

coefficient are derived to first and second derivation where we got 18, 36, and 54 WDCC coefficients. The graphical representation of the extracted WDCC 18, 36, and 54 coefficient is shown in Figs. 9, 10 and 11 respectively.

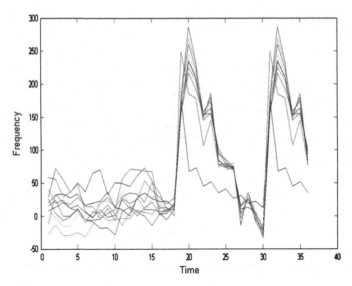

Fig. 10 The WDCC extracted 36 features of word "गणकयंत्र"

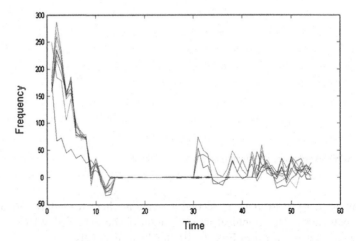

Fig. 11 The WDCC extracted 54 features of word "गणकयंत्र"

4 Design of Marathi Speech Interface System

For the speech recognition system speech recording (database creation), feature extraction, training, classification, and testing are fundamental steps.

4.1 Database Design

The collection of utterances in the proper manner is called the database. We implemented this prototyping application as a speaker dependent. The total number of words with probability 372 utterance is 20, and the data were collected in 03 session so the overall 22,320/- vocabulary size are collected in the database. The sampling frequency for all recordings was 16,000 Hz at the room temperature and normal humidity. The speech data are collected with the help of microphone realtech and matlab software using the single channel. The preprocessing is done with the help of computerized Speech Laboratory (CSL).

4.2 Marathi Speech Activated Calculator (MSAC)

In this research, our objective is to develop MSAC application. Figure 12 describes the basic structural diagram for talking calculator. The voice is recognized and

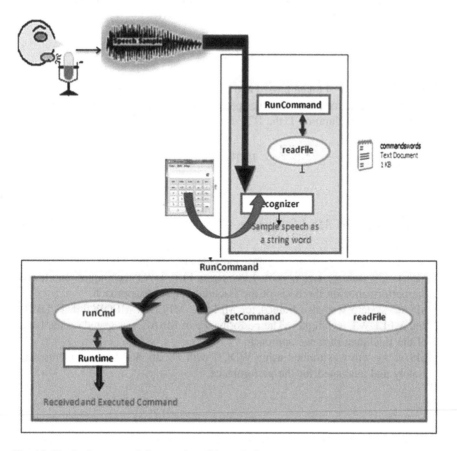

Fig. 12 The basic structural diagram for talking calculator

specific action taken toward voice commands. The MSAC is the speaker-dependent interface system.

5 Experimental Analysis

This application deals with defined set of experiments related to calculator applied on the database designed for this research work.

5.1 Performance of the Marathi Speech Activated Calculator (MSAC) System

The performance of the system is calculated on the basis of accuracy as well as a real time factor. The real-time factor (RTF) is the time required for recognition in response to the operation. The accuracy is calculated on the basis of confusion matrix in which number of token was passed randomly.

$$\text{Accuracy} = \frac{N - C}{N} * 100$$

where N is a number of token passed and C is a number of token confuse. The RTF is a common metric for computing the speed of an ASR system. If it takes time **P** to process an input of duration **I**, the RTF is defined as

$$\text{RTF} = \frac{P}{I}$$

5.2 Training of MSAC System

- MSAC system was trained using individually for the MFCC (13 feature), MFCC (39 feature), LDA, PCA, LPC, Rasta-PLP, DWT techniques, and tested the performance on the basis of the Euclidian distance approach.
- MSAC system was also trained for MFPCA, MFLDA, MFDWT, MFPDWT, and MFLDWT techniques. The performance of MSAC was tested on the basis of the Euclidian distance approach.
- MSAC system was trained using WDCC with 18, 36, and 54 coefficients separately and evaluated for the performance.

5.3 Testing of MSAC System

(a) *Basic Feature Extraction*

In this approach, MSAC system was tested using MFCC (13 feature) and MFCC (39 feature), LDA, PCA, LPC, Rasta-PLP, and DWT techniques. The 13 isolated words are used for testing. The performance of these techniques is calculated on the basis of average accuracy and RTF.

- *MFCC based MSAC performance*
  The performance of MSAC using MFCC is considered on the basis of 18 and 39 coefficients. Total 13 words were tested for 32 trials. The average performance of MFCC for 18 and 39 coefficients are calculated as 75.78 and 78.03, respectively. The RTF for MFCC 18 and 39 coefficients is the 26 and 38 s, respectively.
- *LDA-based MSAC performance*
  Total 13 words were tested for 32 trials. The average performance of LDA is calculated as **67.17 %.** The responding time (RTF) for the action taken is **46 s.**
- *PCA-based MSAC performance*
  Total 13 words were tested for 32 trials. The average performance of PCA is calculated as **62.19 %.** The RTF for MSAC using PCA techniques is **38 s.**
- *LPC-based MSAC performance*
  Total 13 words were tested for 32 trials. The average performance of LPC is calculated as **61.23 %.** The RTF for recognition and action taken in calculator is **51 s.**
- *Rasta-PLP-based MSAC performance*
  Total 13 words were tested for 32 trials. The average performance of Rasta-PLP is calculated as **68.27 %.** The RTF for recognition and action taken in calculator is **48 s.**
- *DWT based MSAC performance*
  Total 13 words were tested for 32 trials. The average performance of Rasta-PLP is calculated as **71.02 %.** The responding time for the calculator for performing the action is **32 s.**

The comparative performance of the MSAC with different feature extraction techniques is described in Table 2.

From Table 2, it is observed efficient accuracy is achieved with MFCC for 39 coefficients but RTF is bit increased than MFCC with 13 coefficients. MFCC 13 coefficient proved to be effective in term of accuracy and RTF.

(b) *Fusion-Based Feature Extraction Techniques*

In the fusion feature extraction techniques base MSAC testing, we have tested the system using MFLDA, MFPCA, MFDWT, MFLPDWT, and MFLDWT techniques. We explored fusion approach for system implementation. If the dimension

Table 2 The performance of MSAC of available feature extraction technique in literature

Sr. no	Technique	Accuracy (%)	RTF (%)
1	PCA	62.19	38
2	LDA	67.17	46
3	LPC	61.23	51
4	Rasta-PLP	68.27	48
5	DWT	71.02	32
6	MFCC (13 feature)	75.78	26
7	MFCC (39 feature)	78.03	38

of the features is reduced without loss of information, this will reduce the RTF and the MFCC provides higher accuracy so we fuse these techniques with MFCC.

- *MFLDA-based MSAC performance*

Total 13 words were tested for 35 trials. The average performance of MFLDA is calculated as **87.90 %.** The responding time for the calculator for performing the action is **22 s.**

- *MFPCA-based MSAC performance*

Total 13 words were tested for 35 trials. The average performance of MFPCA is calculated as **87.12 %.** The responding time for the calculator for performing the action is **20 s.**

- *MFDWT-based MSAC performance*

. Total 13 words were tested for 35 trials. The average performance of MFPCA is calculated as **86.17 %.** The responding time for the calculator for performing the action is **6 s.**

- *MFPDWT-based MSAC performance*

Total 13 words were tested for 35 trials. The average performance of MFPCA is calculated as **89.09 %.** The responding time for the calculator for performing the action is **12 s.**

- *MFLDWT-based MSAC performance*

Total 13 words were tested for 35 trials. The average performance of MFPCA is calculated as **90.12 %.** The responding time for the calculator for performing the action is **10 s.**

The comparative performance of MFLDA, MFPCA, MFDWT, MFPDWT, and MFLDWT are described in Table 3.

The MSAC Application is the real interface application which requires lowest real time factor. From the above results, it is observed that the MFLDWT is proven a higher accuracy than other techniques with acceptable RTF.

Table 3 Performance of the MSAC system using fusion approach

Sr. no	Fusion approach	Accuracy (%)	RTF
1	MFLDA	87.9	22
2	MFPCA	87.12	20
3	MFDWT	86.17	6
4	MFPDWT	89.09	12
5	MFLDWT	90.12	10

In the above fusion approach, we observed that the wavelet play an important role for the reducing the RTF so we proposed our own techniques on the basis of wavelet basic properties known as WDCC.

(c) *WDCC: A proposed approach*

This wavelet decomposed Cepstral Coefficient is used for basic feature extraction. This technique extracted basic feature, first derivative as well as second derivative. From this technique extracted the 18 basic features, 36 features (18 + First Derivative) and 54 features (18 + First derivative + Second derivatives).

The performance of MSAC is tested on the basis of the proposed WDCC approach. The MSAC system is tested using WDCC 18, 36, and 54 coefficients approach. Total 13 words were tested for 35 trials. The average performance of WDCC with 18, 36, and 54 coefficients is calculated as **94.03, 95.01, and 98.04,** respectively. The responding time for the calculator for performing the action using WDCC 18, 36, and 54 coefficients are 9, 20, and 27 s. The comparative performance of the MSAC system using WDCC technique of different features is described in Table 4.

From the above result, the WDCC with 54 coefficients gives the high accuracy with the RTF is slightly greater than other 18 and 36 coefficient. We observe that WDCC with 18 coefficients provided lowest RTF so we tested the MSAC system using WDCC using 18 coefficients.

For this MSAC interface system, we did 6 h training of the dataset. The testing environment varied the system performance. We test the MSAC interface system real time in the air conditioners in office environment, system processor sound, and group talking environment.

Error Rate in Noisy Environment for MSAC System

This MSAC system is the speaker-dependent system. The experiment is tested in noisy environment. The testing environment varied the system performance. We test the MSAC interface system real time in the following noisy environment. The error rate according to the acquisition environment is presented in Table 5.

Table 4 Performance of MSAC system using WDCC approach

Sr. no	Technique	Accuracy (%)	RTF
1	WDCC (18 coefficient)	94.03	9
2	WDCC (36 coefficient)	95.01	20
3	WDCC (54 coefficient)	98.04	27

Table 5 Error rate of MSAC in the different testing environment

Sr. no	Environment	Error rate (%)
1	Air conditioners	13
2	System processor sound	11
3	Group talking environment	10

- **Air conditioners in an office environment** (when the air conditions are started in the office and the sound of the air conditions is mixed with test recorded samples).
- **System Processor sound** (The system processor sound is become maximum and it mixed with the test sample)
- **Group talking environment** (This system tested in real laboratory, number of people talking of each other naturally, they not knowing the system testing background).

5.4 Significance of MSAC

The salient feature of this research as below…

- In the current era of technology, the evolution of speech recognition is done day by day so it is very necessary to bring this technology to the societies in regional language. It is observed that work done in Marathi has not received much more attention. Thus, we have attempted to design and developed Marathi Speech Activated Calculator (MSAC).
- The MSAC system is directly beneficial for society people, where no need of computer literacy and knowledge of English.
- The clustering approach such as PCA, LDA, and LPC was considered for feature extraction, and this gives a new path for speech researcher towards an implementation of real-time application.
- From 1939 to till date, MFCC is one of the robust and dynamic feature extraction techniques; in this experiment, we proposed WDCC feature extraction technique, which gives better performance than MFCC.
- It is very important to adapt research in real-time application; MSAC is the real-time application in Marathi regional language. This will be a chance for Marathwada people to adapt new technology in speech recognition through which they become the part of today's modern technology.

6 Conclusion

From the enrich literature, it is observed that MFCC, PCA, LDA, LPC Rasta-PLP, and much more techniques available for feature extraction. Individual techniques have their own limitation. We tried to come up with these limitations using fusion approach. The database for the said application is recorded by standard protocol.

The MSAC system is tested on the basis of individual feature extraction techniques, fusion approach of MFCC and proposed WDCC approach. From the analysis, we observed the following:

- Efficient accuracy is achieved with MFCC for 39 coefficients but RTF is bit increased than MFCC with 13 coefficients. MFCC 13 coefficient proved to be effective in term of accuracy and RTF.
- MFLDWT is proven a higher accuracy than other fused techniques with acceptable RTF.
- WDCC with 54 coefficients gives the high accuracy but the real time factor is slightly greater than other 18 and 36 coefficients.
- WDCC with 18 coefficients provided acceptable accuracy with lowest RTF.
- The testing environment such as air conditioners in office, system processor sound and group talking environment varies the performance of MSAC.

From the above study the author recommended the WDCC feature extraction techniques are robust and dynamic as compare to MFCC, LPC, Rasta, LDA, and PCA.

References

1. A review on speech recognition technique. Int. J. Comput. Appl. **10**(3), 0975–8887 (2010)
2. Picheny, M.: Large vocabulary speech recognition **35**(4):42–50 (2002)
3. Arokia Raj, A., Susmitha, R.C.: A voice interface for the visually impaired. In: 3rd International Conference: Sciences of Electronic, Technologies of Information and Telecommunications March 27–31, Tunisia (2005)
4. Roux, J.C., Botha, E.C., Du Preez, J.A.: Developing a multilingual telephone based information system in African languages. In: Proceedings of the Second International Language Resources and Evaluation Conference, no. 2, pp. 975–980. ELRA, Athens (2000)
5. Robertson, J., Wong, Y.T., Chung, C., Kim, D.K.: Automatic speech recognition for generalized time based media retrieval and indexing. In: Proceedings of the Sixth ACM International Conference on Multimedia, pp. 241–246. Bristol (1998)
6. Scan soft: Embedded speech solutions. http://www.speechworks.com/ (2004). Accessed 25 Jan 2013
7. Kandasamy, S.: Speech recognition systems. SURPRISE J. **1**(1) (1995)
8. Dusan, S., Rabiner, L.R.: On integrating insights from human speech perception into automatic speech recognition. In: Proceedings of INTERSPEECH 2005. Lisbon (2005)
9. Shrawankar, U., Thakare, V.: Speech user interface for computer based education system. In: International Conference on Signal and Image Processing (ICSIP), pp. 148–152 (2010) (15–17 Dec)
10. Alt, F.L., Rubinoff, M., Yovitts, M.C.: Advances in Computers, pp. 165–230. Academic Press, New York
11. Rebman Jr., C.M., Aiken, M.W., Cegielski, C.G.: Speech Recognition in the Human–Computer Interface, vol. 40, Issue 6, pp. 509–519, Information & Management. Elsevier (2003)
12. Furui, S.: 50 Years of progress in speech and speaker recognition research. ECTI Trans. Comput. Inf. Technol. **1**(2) (2005)

13. Nehe, N.S., Holambe, R.S.: New feature extraction techniques for Marathi digit recognition. Int. J. Recent Trends Eng. **2**(2) (2009)
14. Bhosale, R.S.: Enhanced speech recognition using ADAG SVM approach. Int. J. Emerg. Trends Technol. Comput. Sci. (IJETTCS) **1**(4) (2012)
15. Anumanchipalli, G., Chitturi, R., Joshi, S., Kumar, R., Singh, S.P., Sitaram, R.N.V., Kishore, S.P.: Development of indian language speech databases for large vocabulary speech recognition systems. In: Proceedings of International Conference on Speech and Computer (SPECOM). Patras (2005)
16. Neti, C., Rajput, N., Verma, A.: A large vocabulary continuous speech recognition system for Hindi. In: Proceedings of the National conference on Communications, pp. 366–370. Mumbai (2002)
17. Gawali, B.W., Gaikwad, S., Yannawar, P., Mehrotra, S.C.: Marathi Isolated Word Recognition System using MFCC and DTW Features. ACEEE (2010)
18. Chakraborty, K., Talele, A., Upadhya, S.: Voice recognition using MFCC algorithm. Int. J. Innovative Res. Adv. Eng. (IJIRAE) **1**(10) (2014). ISSN: 2349-2163
19. Patel, K., Prasad, R.K.: Speech recognition and verification using MFCC & VQ. Int. J. Emerg. Sci. Eng. (IJESE) **1**(7) (2013). ISSN: 2319–6378
20. Oh-Wook Kwon, Chan, K., Lee, T.-W.: Speech feature analysis using variational bayesian PCA. IEEE Signal Process. Lett. **10**, 137–140 (2003)
21. Gaikwad, S., Gawali, B., Mehrotra, S.C.: Novel Approach Based Feature Extraction For Marathi Continuous Speech Recognition, pp. 795–804. ACM Digital Library, New York (2012). ISBN: 978-1-4503-1196-0/2012
22. Hermansky, H., Morgan, N.: RASTA processing of speech. IEEE Trans. Speech Audio Process. **2**, 578–589 (1994). doi:10.1109/89.326616
23. Ali, H., Ahmad, N., Zhou, X., Iqbal, K., Muhammad Ali, S.: DWT features performance analysis for automatic speech recognition of Urdu. SpringerPlus **3**:204 (2014) doi:10.1186/2193-1801-3-204
24. Tiwari, A., Zadgaonkar, A.S.: Debauchee's wavelet analysis of speech signal of different speakers for similar speech set. Int. J. Adv. Res. Comput. Sci. Softw. Eng. **4**(8) (2014)
25. Mallat, S.: A Wavelet Tour of Signal Processing. Academic Press (1998)
26. Chan, Y.T.: Wavelet Basics. Kulwer Academic Publications (1995)

Fusion-Based Noisy Image Segmentation Method

Mateusz Buczkowski and Khalid Saeed

Abstract A modified algorithm for segmenting microtomography images is given in this work. The main use of the approach is in visualizing structures and calculating statistical object values. The algorithm uses localized edges to initialise snakes for each object separately then moves curves within the images with the help of gradient vector flow (GVF). This leads to object boundary detection and obtain fully segmented complicated images with the aid of methods like region merging and multilevel thresholding.

Keywords Image segmentation · Canny-Deriche edge detector · Gradient vector flow · Active contour · Bilateral filter

1 Introduction

Images studied in this paper are obtained using microcomputed tomography (μCT) method. Imaging using μCT is a powerful technique for non-destructive internal structure imaging of small objects. Best μCT devices available today can obtain the resolution even better than one micrometer. That advantage lets μCT to be widely used in biology, geology, material science and many other areas where imaging of small structures is required. The general idea behind μCT measurements is to generate electromagnetic radiation with X-ray tube. That radiation after penetrating the sample is deposited in 2D detector on the opposite side of the sample. The detector registers the attenuation of the X-ray intensity. The registered 2D array of

M. Buczkowski (✉)
Faculty of Physics and Applied Computer Science,
AGH University of Science and Technology, Krakow, Poland
e-mail: mateusz.buczkowski1@gmail.com

K. Saeed
Faculty of Computer Science, Bialystok University of Technology, Bialystok, Poland
e-mail: k.saeed@pb.edu.pl

© Springer India 2016
R. Chaki et al. (eds.), *Advanced Computing and Systems for Security*,
Advances in Intelligent Systems and Computing 396,
DOI 10.1007/978-81-322-2653-6_2

21

X-ray intensities is called "projection". Intensity of registered radiation depends on the material radiation absorption property across the single ray. Generally denser materials absorb more radiation. Sample is rotating and few hundred or thousand projections are registered. Computer software is used to reconstruct 3D object from a set of 2D projections using one of the available methods, such as one of the most popular filtered back projection methods based on Radon transform theorem [1]. The 3D object obtained from this step is represented by a 3D matrix of voxels. 2D slices of objects voxels could be represented as 2D image. In this paper images of porous structures are studied. Porous materials can be described as a two-phase composite where one phase is a solid phase and the other is a void or some gas or liquid phase. Separation of these phases by segmenting 2D cross-section images separately is studied in this paper. When differences in X-ray linear attenuation factor for both phases are high, the solution is easier but for images studied in this paper the differences are small in intensities of pixel values for both phases. Studied images have large amount of noise which is also a difficulty. In easy cases, the segmentation could be performed using filtering step such as median or bilateral filter and binarized by simple thresholding method, even with one threshold. This approach could be then applied to all images in stack to obtain all properly seg-mented images. From binary image stack we could render a 3D visualization, for example. Images studied in this paper require a more sophisticated method to obtain fully segmented images.

The work is the extended version of the authors' work in the Second International Doctoral Symposium on Applied Computation and Security Systems organized by University of Calcutta [2]. More details and examples are given in this paper. Some theoretical aspects are repeated for the reader's convenience.

2   Used Methods

2.1   Bilateral Filter

For obtaining properly detected edges from a noisy image, a proper smoothing stage is required prior to edge detection. When processing noisy images this step is crucial for obtaining good results of entire approach. We found bilateral filter is the proper way of smoothing microtomography images presented in this paper. Bilateral filter is a technique which allows to remove unwanted details (textures, noise), and still preserving edges without blurring is the great advantage of this method. Bilateral filter uses a modified version of Gaussian convolution. In Gaussian filtering, weighted average of the adjacent pixels intensities in the given neighbourhood results in new value of the considered pixel. Weights decrease along with the increasing spatial distance from the central pixel (1). Moreover, pixels are less significant for new value of the processed pixel. That dependency is given as

$$G[I]_p = \frac{1}{W_{pG}} \sum_{q \in S} G_\sigma(||p - q||)I_q, \tag{1}$$

where $G_\sigma(x)$ is Gaussian convolution kernel given by (2).

$$G_\sigma(x) = \frac{1}{2\pi\sigma^2} \exp\left(-\frac{x^2}{2\sigma^2}\right) \tag{2}$$

where: S is the spatial domain, W_{pG}—sum of all weights, I—intensity of pixel, $||p - q||$—the Euclidean distance between the considered central pixel p and another pixel q form the given neighbourhood. Profile of weights changes depending on spatial distance as given by σ. Higher sigma results in higher smoothing level. Main disadvantage of Gaussian filter is edge blurring.

Bilateral filter is defined by (3).

$$B[I]_p = \frac{1}{W_{pB}} \sum_{q \in S} G_{\sigma_s}(||p - q||)G_{\sigma_r}(||I_p - I_q||)I_q, \tag{3}$$

where

$$W_{pB} = \sum_{q \in S} G_{\sigma_s}(||p - q||)G_{\sigma_r}(||I_p - I_q||). \tag{4}$$

Only pixels close in space and intensity range are considered (close to the central pixel). Spatial domain Gaussian kernel is given by G_{σ_s} Weights decrease with increasing distance. Range domain Gaussian kernel is given by G_{σ_r}. Weights decreases with increasing intensity distance. Simultaneous filtering in both spatial and intensity domain gives bilateral filter capability of smoothing image (background and object area) and preserve edges at the same time [3]. That kind of behaviour is crucial when processing noisy images demanding high level of smoothing to remove noise.

2.2 Canny–Deriche Edge Detector

Canny formulated three important criteria for effective edge detection in his paper [4]:

- Good detection—low probability of failing to detect existing edges and low probability of false detection of edges
- Good localization—detected edges should be as close as possible to the true edges

- One response to one edge—multiply responses to one real edge should not appear

Canny combined these criteria into one optimal operator (approximately the first derivative of Gaussian) [4, 5]—see (5).

$$f(x) = -\frac{x}{\sigma^2} e^{-\frac{x^2}{2} \cdot \sigma^2}. \tag{5}$$

Deriche modified Canny's approach to obtain a better optimal edge detector [5]. He presented his optimal edge detector in the form of:

$$f(x) = k \cdot e^{-\alpha \cdot |x|} \sin \omega x \tag{6}$$

and for the case when ω tends to 0

$$g(x) = k \cdot x e^{-\alpha \cdot |x|}. \tag{7}$$

Performance of that approach is better than Canny's original idea. At the beginning calculation of magnitude and gradient direction are performed to obtain gradient map. Higher gradient values are obtained near to the edges of objects. Then non-maximal suppression selects the single brightest pixel across the width of an edge which is a thin edge. Last stage involves hysteresis thresholding performed to get the final result of edge detection. Hysteresis thresholding uses two thresholds as parameters. Accordingly thresholds pixels are divided into three groups. Pixels with values below low threshold are removed which means that they are classified as non-edges. Pixels with values above high threshold are retained so they are considered as edges. Pixel with intensity value between low and high threshold is considered as edge pixel only if connected to some pixel above high threshold [4, 5].

2.3 Active Contours (Snakes) and Gradient Vector Flow (GVF)

Snake could be described as parametric curve

$$x(s) = [x(s), y(s)], \quad s \in [0, 1]. \tag{8}$$

That snake could move in spatial domain of the image to minimize energy functional

$$E = \int_0^1 \frac{1}{2} \left[\alpha |x'(s)|^2 + \beta |x''(s)|^2 \right] + E_{ext}(x(s)) \mathrm{d}s. \tag{9}$$

where α and β are weighting parameters controlling tension (first derivative) and rigidity (second derivative). E_{ext} is obtained from image gradient map. It takes smaller values near objects of interest such as edges. In our approach this external force obtained from gradient vector flow (GVF) method is computed as a diffusion of the gradient vectors. GVF method could be applied for example to a grey-level or a binary edge map derived from the image. GVF fields are dense vector fields derived from images by minimizing energy functional. The minimization is achieved by solving a pair of decoupled linear partial differential equations that diffuses the gradient vectors of edge map obtained from the image. Active contour using GVF field as external force could be named GVF snake. Detailed description and numerical implementation could be found in original GVF paper [6].

2.4 Statistical Region Merging (SRM) and Multilevel Thresholding

In region merging-based method, regions are described as sets of pixels with homogeneous properties and they are iteratively grown by combining smaller regions. Pixels are elementary regions. Statistical test is performed to decide if merge tested regions. Detailed description is available in original SRM paper [7]. Multilevel thresholding modify Otsu method allowing to get more than two pixel classes by choosing the optimal thresholds by maximizing a modified between-class variance. In this paper pixels were divided into three classes (background, object, holes inside objects together with objects shadows). Detailed description is available in original multilevel thresholding paper [8].

3 The Proposed Methodology

Images analysed in this paper are quite complicated to segment. The major difficulty is that the objects can have intensities of pixels very similar to the background. Sometimes even humans cannot say where exactly object edge is placed. To obtain proper segmentation of these images complex approach combining several methods is required. Simple segmentation methods based on pixel intensity like thresholding do not apply here because they are not good with segmenting noisy images with non-uniform objects [9–13]. This paper focuses on combining various methods like: Canny–Deriche edge detection [5], bilateral filtering [3], gradient vector flow [6], active bontour [6], statistical region merging [7] and multilevel thresholding based on Otsu method [8] to obtain multistage approach with good segmentation results.

In this paper data in the form of 8-bit grayscale images were used. First histogram normalization is applied to the original image. Due to high noise level of images, efficient smoothing is crucial to make proper segmentation possible.

Smoothing step is based on bilateral filter. Adjustable Parameters for bilateral filter are: mask size, intensity range, spatial sigma and intensity sigma. Finding proper values of these parameters using semi-automatic approach was described in the authors' previous work [14]. When proper values of the parameters are obtained for one image from bigger set those values could be used to process entire μCT images set. Smoothing step was iteratively applied twice, once with bigger parameters values (more smoothing) and then with smaller parameters values (less smoothing, see Fig. 1). This step removes significant amount of noise and unwanted textures resulting in the simplified image.

In the edge detection stage Canny–Deriche edge detector is applied. Adjustable Parameters for edge detection step are: alpha, high threshold, low threshold. Finding proper values of these parameters were described in previous article [14]. In this step, gradient magnitude and direction are calculated then non-maximum suppression is performed which allows thin edges (Fig. 2a, b). Hysteresis thresholding is performed at the end of process to obtain most relevant edges (Fig. 2c).

Object grouping is performed to group all edges into array of objects constructed from edges. This is achieved by grouping all edge pixels which are close to each other within chosen radius. From that point all objects are processed separately. Objects are processed with GVF method (Fig. 3) to obtain gradient map proper for active contour method. Snake is initialized outside each object and evolves to find object boundaries by minimalizing energy of snake at each iteration (Figs. 4 and 5). Energies used: internal energy (first and second derivative), external energy (obtained from gradient map) and external pressure force. Parameters were chosen to perform well on this kind of images. In our approach snake is discretized. Finite number of control points was used to calculate total snake energy. For each iteration, snake was resampled to assure proper behaviour. Viterbi algorithm helps with optimization of the contour evolution.

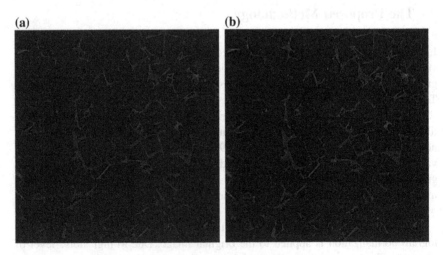

(a) **(b)**

Fig. 1 Bilateral filtering: first iteration (**a**) and second iteration (**b**)

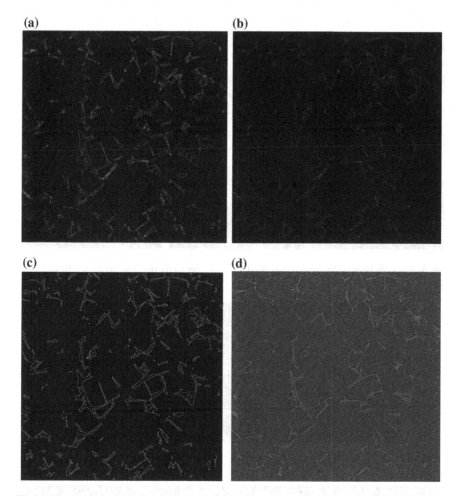

Fig. 2 Canny–Deriche edge detector: gradient map (**a**), non-maximum suppression (**b**), final edge detection result after hysteresis thresholding (**c**) and original image (**d**)

Snake curve allows to create mask which in combination with original object coordinates allows to cut off pixels inside this snake from original image after smoothing (Fig. 6).

Statistical region merging is performed to merge pixels into regions of similar intensity to simplify image (Fig. 7).

Simplified images are finally segmented using multilevel thresholding based on Otsu method (Fig. 8). If objects are smaller than 60 px (width or height) or all pixels in image are higher than some threshold, simple thresholding with one threshold is used. Algorithm flow is shown in Fig. 9.

Finally all binary objects images are combined to obtain a fully segmented input image (Fig. 10).

(a)   (b)   (c)

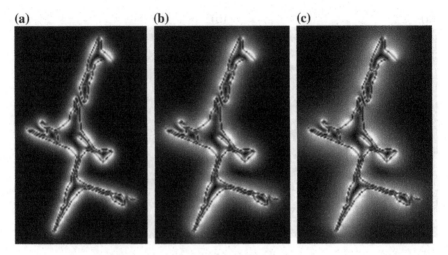

Fig. 3 Example of gradient vector flow field after 100 iterations (**a**), 300 iterations (**b**) and 700 iterations (**c**)

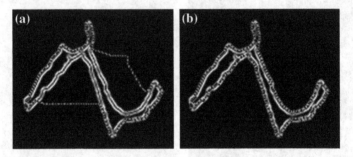

Fig. 4 Example of snake evolution after 15 iterations (**a**) and final snake (**b**)

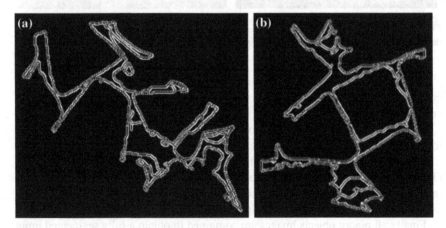

Fig. 5 Two examples of gradient map obtained with GVF method and snake evolved to find object boundaries on top of it (*blue pixels* are discrete points used to calculate snake energy)

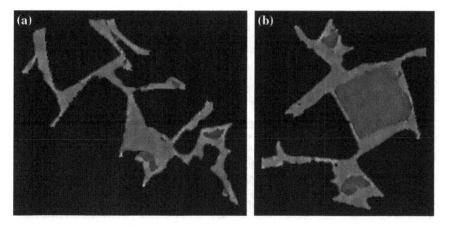

Fig. 6 Two examples of pixels cut off from original image after smoothing with use of mask obtained from snake curve coordinates

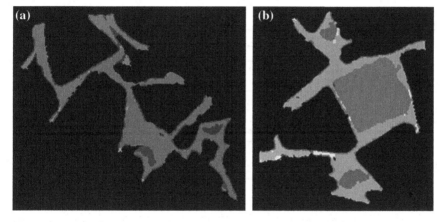

Fig. 7 Two examples of pixels cut off from original image after smoothing with use of mask obtained from snake curve coordinates and after use of statistical region merging method

4   Experimental Results and Interpretation

Presented method allows to treat weak contrast images of porous structures and images containing separate objects giving good results of image segmentation. Further enhancements will be applied in the future to improve results of segmentation. Proper filtering using two-step bilateral filter prior to Canny–Deriche edge detection allows to obtain images with proper localized edges with very small amount of false edges detected and true edges omitted. Some small gaps in edges occurred after edge detection step which was resolved by using active contour

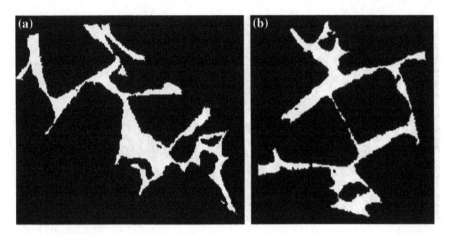

Fig. 8 Two examples of binary images obtained after use of multilevel thresholding based on Otsu method

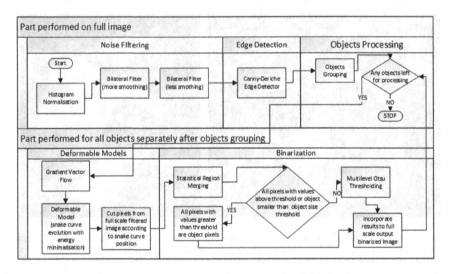

Fig. 9 Flowchart of proposed approach

method along with GVF. Full segmentation of processed image fragments was achieved by using statistical region merging and multilevel thresholding based on Otsu method. All binarized objects were combined into a fully segmented image (Fig. 10). The approach introduced in this paper expands the idea presented in the authors' previous papers [14, 15] which concerned only edge detection. Much better results are achieved with this upgraded approach resulting in fully segmented binary images. Presented approach leaves possibility for future upgrades to obtain even better results.

Fig. 10 Fully segmented image with bigger objects (**a**) and smaller objects (**b**)

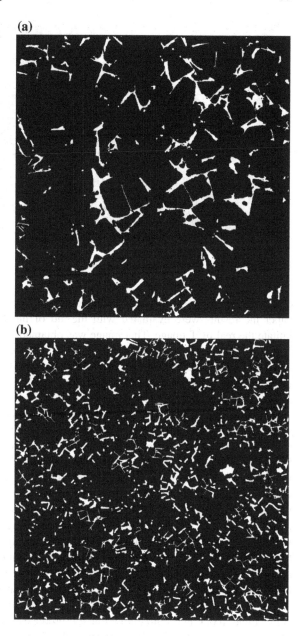

(a)

(b)

5 Evaluation and Comparison of Results

To evaluate the results we need to know the exact position of all object and background pixels. Evaluating results obtained from that kind of images is a difficult task. Marking object contour by hand is time consuming and not so precise

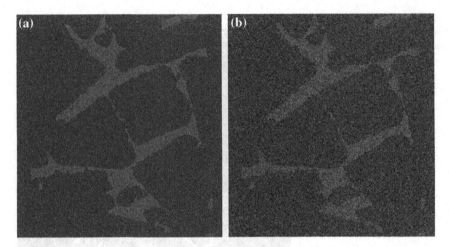

Fig. 11 Fragments of mock test images with noise standard deviation equals to: 12 (**a**), 20 (**b**)

especially when complex object with non-trivial shapes are considered. In this paper the results were evaluated with the aid of mock input image prepared to imitate the real μCT image. That image was obtained from the algorithm with final output as binarized image. Objects and background grayscale level were set according to the average values of those elements obtained from the original μCT images. Then noise with various standard deviations was added to make mock images similar to the real data obtained from μCT device (Fig. 11).

Artificial mock image, with added standard deviation of noise equals to 12 which is similar value to the original one, was prepared. Image was processed using our algorithm noise filtering part (see Fig. 9) combined with single threshold binarization (Fig. 12a), using multilevel Otsu thresholding after the algorithm noise filtering part (Fig. 12c), using multilevel Otsu thresholding after filtering stage without second bilateral filter (Fig. 12b) and using the full approach presented in this paper (Fig. 12d). Several statistical evaluation measures of binary classification for our algorithm such as sensitivity, specificity, precision, negative predictive value and accuracy were presented in Table 1. Accordingly, the results are good and they become even better after some further modification in the last stage of the algorithm after a much precise binarization at this step. Test performed with artificial mock image shows that the approach with simple thresholding gives weak results whilst the approach with multilevel Otsu thresholding produces results comparable to our algorithm output but still produces little noise artefacts (Fig. 12). The same methodology used to binarize real μCT image, shows that the algorithm presented in this paper performs much better with original μCT images than the compared approaches (Fig. 13). Original μCT images are much harder to binarize than artificial mock images introduced to evaluate results. Real images have much more complex structure of objects, such as shadows near borders and various grayscale levels inside objects, sometimes with intensities similar to the background

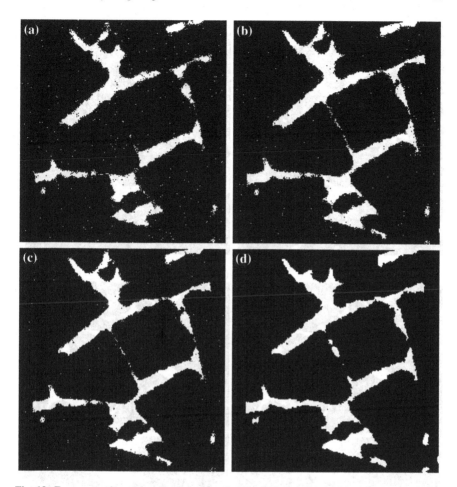

Fig. 12 Fragments of mock test images with noise standard deviation equals to 12, processed by: **a** authors' algorithm filtering part with single threshold binarization, **b** filtering part without second bilateral filter and with multilevel Otsu thresholding, **c** filtering part with multilevel Otsu thresholding, **d** final result of authors' algorithm

Table 1 Segmentation evaluation test results

Noise standard deviation	6	8	10	12	20
Sensitivity	0.964	0.955	0.944	0.932	0.711
Specificity	0.990	0.989	0.980	0.983	0.978
Precision	0.887	0.874	0.803	0.816	0.729
Negative predictive value	0.997	0.996	0.995	0.994	0.976
Accuracy	0.988	0.986	0.978	0.979	0.958

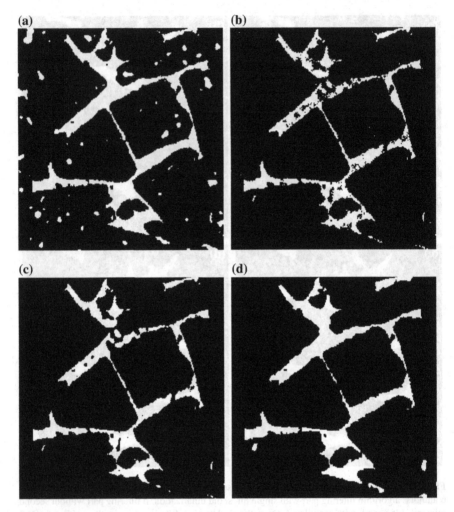

Fig. 13 Fragments of original input image processed by authors' algorithm: **a** filtering part with single threshold binarization, **b** algorithm filtering part without second bilateral filter and with multilevel Otsu thresholding, **c** filtering part with multilevel Otsu thresholding, **d** final result of authors' algorithm

level. Noise also seems to have much more complex structure than simple random noise with given standard deviation. Other tested methods fail because they could not produce uniform and noise-free objects.

Acknowledgements The research was partially supported by doctoral scholarship IUVENES—KNOW, AGH University of Science and Technology in Krakow and by The Rector of Bialystok University of Technology in Bialystok, grant number S/WI/1/2013.

References

1. Radon, J.: Uber die Bestimmung von Funktionen durch ihre Integralwerte Langs Gewisser Mannigfaltigkeiten. Ber. Saechsische Akad. Wiss. **29**, 262 (1917)
2. Buczkowski, M., Saeed, K.: A multistage approach for noisy micro-tomography images. In: ACSS 2015—2nd International Doctoral Symposium on Applied Computation and Security Systems organized by University of Calcutta (2015)
3. Paris, S., Kornprobst, P., Tumblin, J., Durand, F.: Bilateral filtering: theory and applications. Found. Trends Comput. Graph. Vis. **4**(1), 1–73 (2008)
4. Canny, J.: A computational approach to edge detection. IEEE Trans. Pattern Anal. Mach. Intell. **6**, 679–698 (1986)
5. Deriche, R.: Using Canny's criteria to derive a recursively implemented optimal edge detector. Int. J. Comput. Vis. **1**(2), 167–187 (1987)
6. Xu, C., Prince, J.L.: Snakes, shapes, and gradient vector flow. IEEE Trans. Image Process. **7**(3), 359–369 (1998)
7. Nock, R., Nielsen, F.: Statistical region merging. IEEE Trans. Pattern Anal. Mach. Intell. **26**(11), 1452–1458 (2004)
8. Liao, P.-S., Chen, T.-S., Chung, P.-C.: A fast algorithm for multilevel thresholding. J. Inf. Sci. Eng. **17**(5), 713–727 (2001)
9. Chenyang, X., Pham, D.L., Prince, J.L.: Image segmentation using deformable models. Handbook Med. Imaging **2**, 129–174 (2000)
10. He, L., et al.: A comparative study of deformable contour methods on medical image segmentation. Image Vis. Comput. **26**(2), 141–163 (2008)
11. Rogowska, J.: Overview and fundamentals of medical image segmentation. In: Handbook of Medical Imaging, pp. 69–85. Academic Press Inc. (2000)
12. Jahne, B.: Digital Image Processing: Concept, Algorithms, and Scientific Applications. Springer, New York (1997)
13. Sezgin, M., Sankur, B.: Survey over image thresholding techniques and quantitative performance evaluation. J. Electron. Imaging **13**(1), 146–165 (2004)
14. Buczkowski, M., Saeed, K.: A multistep approach for micro tomography obtained medical image. J. Med. Inf. Technol. **23/2014** (2014). ISSN 1642-6037
15. Buczkowski, M., Saeed, K., Tarasiuk, J., Wroński, S., Kosior, J.: An approach for micro-tomography obtained medical image segmentation. In: Chaki, R., et al. (eds.) Applied Computation and Security Systems, Advances in Intelligent Systems and Computing, vol. 304 (2015)

An Approach for Automatic Indic Script Identification from Handwritten Document Images

Sk. Md. Obaidullah, Chayan Halder, Nibaran Das and Kaushik Roy

Abstract Script identification from document images has received considerable attention from the researchers since couple of years. In this paper, an approach for *HSI* (*Handwritten Script Identification*) from document images written by any one of the eight Indic scripts is proposed. A dataset of 782 Line-level handwritten document images are collected with almost equal distribution of each script type. The average Eight-script and Bi-script identification rate has been found to be 95.7 % and 98.51 %, respectively.

Keywords Document image analysis · Handwritten script identification · Multi-classifier · Document fractal dimension · Directional morphological kernel · Interpolation

1 Introduction

One of the important area of research under document image processing is optical character recognition or in short OCR. First, physical documents are digitized by camera, scanner, etc. devices, and then textual information is generated from them

Sk.Md. Obaidullah (✉)
Department of Computer Science & Engineering, Aliah University, Kolkata, West Bengal, India
e-mail: sk.obaidullah@gmail.com

C. Halder · K. Roy
Department of Computer Science, West Bengal State University, Barasat, West Bengal, India
e-mail: chayan.halderz@gmail.com

K. Roy
e-mail: kaushik.mrg@gmail.com

N. Das
Department of Computer Science & Engineering, Jadavpur University, Kolkata, West Bengal, India
e-mail: nibaran@gmail.com

© Springer India 2016
R. Chaki et al. (eds.), *Advanced Computing and Systems for Security*,
Advances in Intelligent Systems and Computing 396,
DOI 10.1007/978-81-322-2653-6_3

by applying OCR techniques. Document digitization and text conversion has its usefulness for better indexing and retrieval of huge volume of data which is available in our modern society. But the problem of OCR become complex due to multilingual and multi-script nature of a country like India, where 22 official languages are present and 13 different scripts are used to write them [1, 2]. Including English which is a very popular language in India the total number of languages increases to 23. In our daily life we come across various documents which are multi-script in nature. Postal documents, preprinted application form, etc. are good example of such documents. To process these documents automatically, we need to design a general class OCR system which will be able to cater all class of scripts. Another solution is to design a script identification system which will identify the nature of the script first, then supply those scripts to the script specific OCR. The feasibility criterion of the former solution is not realistic due to larger number of languages and scripts in India, so we try to solve the problem in the light of second idea. In this scenario the need of an automatic script identification system has become an essential.

Following Fig. 1 shows a block diagram of a script identification system in Indian scenario. Multi-script (both single document written by single script/multiple scripts) is supplied to the system, followed by preprocessing, feature extraction, classification. Finally, specific script type is produced as an output. Afterwards script specific OCR can be called.

- *Previous Work*

The whole work of script identification can be classified into two main categories namely *PSI* (*Printed Script Identification*) or *HSI* (*Handwritten Script Identification*) problem based on type of the document acquired (machine generated or human written). The problem of *HSI* is more challenging than *PSI* due to dynamic nature of writing, i.e., versatility of writing style, variation in interline, interword spacing, character sizes from different writers across the globe. In the

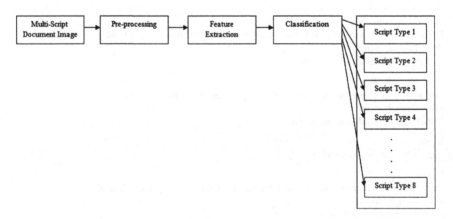

Fig. 1 Block diagram of the proposed multi-script document processing system

literature, many works are reported related to *PSI* and *HSI* on Indic scripts. Among those set of works, few of the *PSI* techniques are depicted in [3–8]. To talk about *HSI* category, a scheme was proposed by Hochberg et al. [9] to identify six Indic and non-Indic scripts namely Arabic, Chinese, Cyrillic, Devanagari, Japanese and Latin using some features like sphericity, aspect ratio, white holes, etc. Another technique was proposed by Zhou et al. [10] to identify Bangla and English scripts using connected component profile-based features. Singhal et al. [11] proposed an approach to identify Roman, Devanagari, Bangla, and Telugu scripts from line-level handwritten document images. They used rotation invariant texture features based on multichannel Gabor filtering and Gray level co-occurrence matrix as principal feature set. Roy et al. [12] proposed a technique to identify Bangla and Roman scripts for Indian postal automation using the concept of water reservoir and busy zone. Hangarge et al. [13] identified Roman, Devanagari, and Urdu script using a texture-based algorithm. The work was done at the block level using some visual discriminating features. In a recent work, the same author [14] proposed a word-level directional DCT-based approach to identify six different Indic scripts. In another very recent work, Pardeshi et al. [15] proposed a scheme for word-level handwritten script identification from 11 Indic scripts using transform-based features like discrete cosine transform, radon transform, etc. But in terms of different performance matrices *HSI* techniques are still lagging far behind the PSI techniques developed so far. That is why *HSI* on Indic scripts is still an open challenge.

In this paper, we propose an *HSI* technique to identify eight handwritten Indic scripts namely *Bangla, Devnagari, Kannada, Malayalam, Oriya, Roman, Telugu,* and *Urdu*. A multidimensional feature set is constructed by observing different properties of these eight scripts. The paper is organized as follows: In Sect. 2 proposed methodologies are described which includes preprocessing and feature extraction. Section 3 provides experimental details where dataset preparation, experimental protocol, evaluation methodologies, results, and analysis are discussed. Conclusion and future scopes are discussed in Sect. 4. Finally, references of the literature are mentioned in the last section.

2 Proposed Methodology

2.1 Preprocessing

Data collected from different sources are initially stored as gray scale images. A two-stage-based binarization algorithm [12] is used to convert these 256-level grayscale images to two-tone binary images. At first stage local window-based algorithm is applied to get information about different ROI (Region Of Interest). RLSA (Run Length Smoothing Algorithm) is applied afterwards on those pre-binarized images to reduce presence of stray or hollow regions. Then using component labeling each component obtained after the first stage is selected and mapped into the original gray scale image. Final version of the binarized image is obtained by applying a global binarization algorithm on each of these regions. This two-stage-based technique has advantage that the binarized image will be at least as good as if only global thresholding method would have been applied. After preprocessing feature extraction is carried out to generate multidimensional feature set. Following section discusses the major features used for the present work.

2.2 Feature Extraction

One of the most important tasks in any pattern recognition work is to collect 'proper' feature set. Here, by the term 'proper' we mean the set of features which are robust enough to capture maximum interscript variability, while obtain minimum intrascript variations. These features should be computationally easy and fast also. Following section provides a glimpses of the important features used for the present work.

- *Shape or Structure based feature*

Shape or structure of the graphemes of different scripts is a very useful feature on the overall visual appearance of the particular script. We have computed different structural features like convex hull, circularity, rectangularity, etc. on the input images at component level. Following Fig. 2 shows few sample output images: (a) Convex hull drawn on the roman script, (b) Inner and outer circle drawn on the Urdu image component. Here maximum circularity will be obtained when the difference between two radii will be zero. Our observation is Oriya script graphemes are maximum circular nature in compared to others. (c) Rectangular box is drawn on the Roman script component. From each of these structure or shape drawn on different image components we calculate some feature values like convexity distances, maximum, minimum length, their average value, ratios, variance, standard deviation, etc.

- *DFD (Document Fractal Dimension)*

Another important topological feature which is based on the pixel distribution of upper and lower part of the image component has been introduced here. This

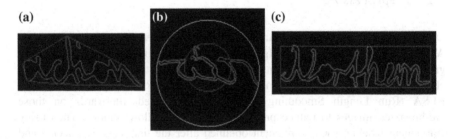

Fig. 2 Computation of structural features (*blue* minimum encapsulating and *red* best fitted)

feature is named here as *Document Fractal Dimension* or in short *DFD. DFD* feature is motivated by the concept of Mandelbrot's fractal geometry theory [17]. A fractal is defined as a set for which the Hausdorff-Besikovich dimension is strictly larger than the topological dimension. The dimension of the fractal is an important property because it contains information about their geometric structure at pixel level. For present work the fractal dimension of the upper part and lower part of each script components has been calculated. Box counting algorithm has been followed where size of a box is assumed to be a unit pixel. A significant role is played by these top and bottom potion of an image component to qualify as a distinguishing feature among 'matra' and non-'matra'-based scripts. For example, Bangla, Devnagari, etc. scripts contains 'matra,' which is a collection of continuous pixel at the top portion of each word or line. Whereas Urdu, Roman, etc. scripts are example of non-'matra'-based scripts. Now if ration of the pixel density is computed for these two cases then there will be a significant difference between these two categories. Figure 3 shows example of *DFD* obtained from each script images. (a) Sample word of original script, (b) *DFD* of the upper part of the contour, (c) *DFD* of the lower part of the contour.

Fig. 3 Fractal dimension **a** original component **b** *upper-fractal* upper part of the contour as fractal **c** *lower-fractal* lower part of the contour (customized word-level outputs are shown)

- *DMK (Directional Morphological Kernel)*

Important morphological operations considered for the present work are dilation, erosion, opening, closing, top-hat, and black-hat transforms. But novelty of the present work is: based on our visual observation of different directional strokes presence in different Indic scripts *Directional Morphological Kernel* or *DMK* has been defined. Four kernels namely *H-kernel, V-kernel, RD-kernel,* and *LD-kernel* are defined. They are 3×11, 11×3, 11×11, and 11×11 matrices correspondingly where horizontal, vertical, right diagonal, and left diagonal pixels are 1 and rests are 0. These four kernels are capable enough to capture the presence of different directional strokes in eight different angles namely 0°, 45°, 90°, 135°, 180°, 225°, 270°, 315°, and 360°. Using these kernels, we computed different morphological transformational operations. Initially original image is dilated using default kernel of OpenCV. The dilated image is then eroded four times using four different kernels (*H-kernel, V-kernel, RD-kernel,* and *LD-kernel*). The ratio of those eroded images with the dilated image is obtained. The average and standard deviation of the eroded images are also computed. Similar kinds of operations are followed for opening, closing, top-hat, and black-hat transformations also.

- *Interpolation based feature*

Image upsize and downsize operation can be performed using interpolation. This simple property has been successfully employed as a useful feature extractor for the present work. Initially image dilation is performed using default 3×3 kernel [18]. Then, the images are interpolated using different mechanism namely nearest neighbor, bilinear, pixel area resampling method, bicubic interpolation. Normally nearest neighbor interpolation takes the closest pixel value for resizing calculation. The 2×2 surroundings are taken for bilinear operation. The virtual overlapping between the resized image and original image is performed and then the average of the covered pixel values is computed in case of pixel area re-sampling method. For bicubic operation a cubic spline between the 4-by-4 surrounding pixels in the source image is fitted then reading off the corresponding destination value from the fitted spline is performed.

- *Feature inspired by Gabor filter*

It is a convolution-based technique used widely for texture analysis. It is one of the most popular band pass filter. It has been observed that the frequency response of Gabor filter is similar to human visual system. The response of Gabor filter to an image is determined by the 2-D convolution operation. In general, the filter will convolve with the input image signal and a Gabor space is generated. If $I(x, y)$ is an image and $G(x, y, f, \phi)$ is the response of a Gabor filter with frequency f and orientation ϕ to an image on the (x, y) spatial coordinate of the image plane [19, 22].

$$G(x, y, f, \phi) = \iint I(p, q) g(x - p, y - q, f, \phi) \, dp dq \qquad (1)$$

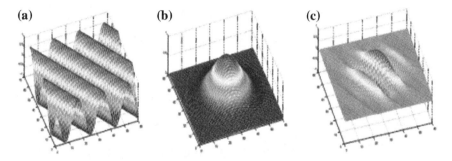

Fig. 4 2-D Gabor filter, **a** a Sinusoid **b** a Gaussian kernel **c** corresponding Gabor filter [19]

Figure 4 shows the 2-D representation of a Gabor filter.

In the proposed approach, multiple feature values are computed forming a Gabor filter bank. The texture variations of different Indic scripts considered for the present work are analyzed. Experimentally, we set the filter with frequency 0.25 and orientation of 60°, 90°, 120°, and 150° for computations of varying Gabor filter inspired features. Afterwards the standard deviation of the real part and imaginary part are computed and considered as feature values.

3 Experimental Details

3.1 Dataset Development

The most time consuming and tedious task for any experimental work is data collection. Availability of benchmark dataset is a problem in this kind of work. Though few works are going on by different researchers on Indic script identification problem but till date no standard handwritten dataset of all official Indic scripts is made available. We collected document image dataset from different persons with varying sex, age, educational qualification, etc. to incorporate maximum variability and realness within the data. For Kannada script we have used the available KHTD [20] handwritten dataset. Lines are extracted from those document pages using a semi-automated technique [12]. Special care was taken to handle touching or skewed handwritten text lines. Finally, a Line-level dataset of total 782 document images with a distribution of 100 Bangla, 100 Devnagari, 102 Kannada, 100 Malayalam, 100 Oriya, 90 Roman, 90 Telugu, and 100 Urdu images are prepared (sample shown in Fig. 5). Document digitization was done using HP flatbed scanner and stored initially at 300 dpi. Binarization was done using existing two-stage-based algorithm that was discussed already. Finally, experimentation was carried out on the prepared dataset (sample shown in Fig. 5).

Fig. 5 Sample line-level document images from our prepared dataset. (*top to bottom*) Bangla, Devnagari, Kannada, Malayalam, Oriya, Roman, Telugu, and Urdu

3.2   Experimental Protocol

The training phase of any classification technique initiates the learning process of distinguishable properties for each of the target script class. During the test phase the dissimilarity measure of the script classes are evaluated. Generation of training and test set data is very crucial decision for any classification scheme. For present work whole data set is divided into training and test sets in equal ratio, i.e., 1:1 ratio. Following section describes about the outcome of the test phase.

3.3   Evaluation Using Multiple Classifiers

Evaluation process is carried out using MLP classifier which we have implemented for experimentation. Simultaneously performance of the proposed technique is also evaluated in multi-classifier environment [21] to observe the robustness of our system. Table 1 shows performances of different classifiers on the present dataset. Six different classifiers namely MLP, logistic model tree, simple logistic, LIBLINEAR, RBFNetwork, and BeyesNet are used here with customized tuning.

Table 1 Statistical performance analysis of different classifiers for eight-script combination

Classifier versus parameter	AAR (%)	MBT (s)	TP rate	FP rate	Precision	Recall	F-measure
MLP	95.7	202.5	0.957	0.006	0.958	0.957	0.957
LMT	94.9	61.36	0.949	0.007	0.949	0.949	0.949
Simple logistic	94.9	17.22	0.949	0.007	0.949	0.949	0.949
LIBLINEAR	90.1	4.79	0.900	0.014	0.900	0.900	0.901
RBFNetwork	88.3	8.55	0.882	0.017	0.882	0.882	0.884
BayesNet	86.7	0.28	0.867	0.019	0.867	0.867	0.868

Seven evaluation matrices namely AAR (Average Identification Rate), MBT (Model Building Time), TP rate (True Positive Rate), FP rate (False Positive Rate), precision, recall and F-measure are evaluated. Detail information about these classifiers and evaluating matrices are available from the work of Obaidullah et al. [16]. Experimental results shows effectiveness of MLP classifier, which obtain highest Eight-script average identification rate of **95.7 %**, followed by LMT and simple logistic both 94.9 %, LIBLINEAR 90.1 %, RBFNetwork 88.3 %, and BayesNet 86.7 %. In terms of MBT, BayesNet converges very fast among all and MLP takes maximum time to build the model on present dataset. A tradeoff between AAR and MBT need to be chosen while selecting appropriate classifier in real life scenario.

Following Fig. 6 shows a sample diagram of a MLP with input, hidden and output layers. In present work, the configuration of MLP is 148-34-8 as the number of neurons in input and output layers are 148 and 8, respectively. The number of neurons in hidden layer is calculated by a heuristic formula. The experimentation was carried out for an epoch size of 500.

3.4 Result and Analysis

Table 2 shows the confusion matrix using MLP on the test dataset. Three and two Bangla scripts images are misclassified with Malayalam and Telugu corresponding-ingly. For Devnagari, total three images are misclassified, out of which one with Bangla and two with Malayalam. Similar kind of few misclassified instances can be found for other scripts also. It has been observed that scripts like Urdu whose characters/graphemes are unique in nature compared to other Indic scripts has successfully identified and no misclassification has been found. We have deeply observed the misclassification patterns and found that, this misclassification occurs due to dynamic change of handwriting of different writes at different time instances. Structural similarity is another important issue that has been found as a reason of misclassification.

A graph is shown in Fig. 5 comparing the average identification rate of different classifiers. MLP obtained highest average identification rate and others appear in

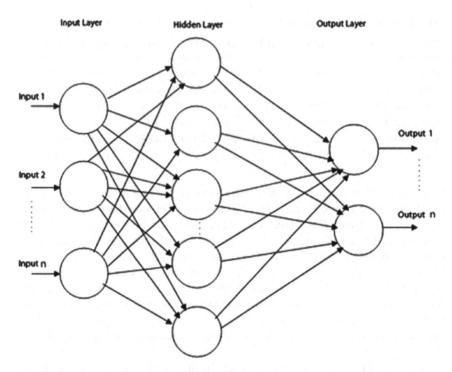

Fig. 6 Sample diagram of MLP neural network with input, hidden, and output layers

Table 2 Confusion matrix using MLP classifier on the test dataset

Script name	BEN	DEV	KAN	MAL	ORY	ROM	TEL	URD
BEN	44	0	0	3	0	0	2	0
DEV	1	47	0	2	0	0	0	0
KAN	0	0	45	2	0	1	1	0
MAL	0	2	1	50	0	0	0	0
ORY	0	0	0	0	41	0	0	0
ROM	0	0	0	0	0	42	0	0
TEL	0	0	0	0	0	2	53	0
URD	0	0	0	0	0	0	0	52

Average eight-script identification rate using MLP: 95.7 %

very near proximity. This performance graph justifies the robustness of the feature set implemented for the present work (Fig. 7).

Table 3 shows the average Bi-script classification rate using MLP. In introductory section we have mentioned that multi-script documents are in general two types. One is single document written by single script and another is single document written by multiple scripts. The all-script (here Eight-script) average identification rate is suitable evaluating parameter for the former case (single document

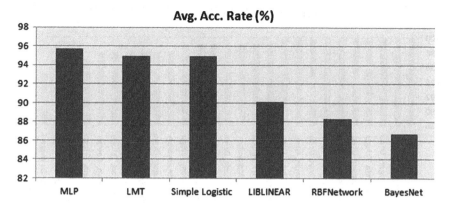

Fig. 7 Performance comparison of different classifiers

Table 3 Bi-script identification rate using MLP classifier

Sl. No.	Script combination	AAR (%)	Sl. No.	Script combination	AAR (%)
1	Bangla, Urdu	100	15	Bangla, Oriya	99
2	Devnagari, Roman	100	16	Malayalam, Oriya	99
3	Devnagari, Urdu	100	17	Roman, Telugu	98.9
4	Kannada, Telugu	100	18	Bangla, Malayalam	98
5	Kannada, Urdu	100	19	Devnagari, Oriya	98
6	Malayalam, Telugu	100	20	Malayalam, Urdu	98
7	Oriya, Roman	100	21	Bangla, Roman	97.9
8	Oriya, Urdu	100	22	Devnagari, Telugu	97.9
9	Roman, Urdu	100	23	Malayalam, Roman	97.9
10	Telugu, Urdu	100	24	Bangla, Telugu	96.9
11	Bangla, Kannada	99.01	25	Kannada, Roman	96.9
12	Devnagari, Kannada	99.01	26	Oriya, Telugu	96.9
13	Kannada, Malayalam	99.01	27	Bangla, Devnagari	95
14	Kannada, Oriya	99.01	28	Devnagari, Malayalam	92

Average Bi-script identification rate: 98.51 %

written by single script) and Bi-script average identification rate is truly justified for the later one (single document written by multiple scripts). That is why we have thoroughly experimented $^8C_2$ or 28 Bi-script combinations and **98.51** % average identification rate is found for the same. Among all, 100 % identification rate is obtained by 10 combinations. Total 26 combinations shows higher identification rate compared to Eight-script average identification rate. Only two instances namely Bangla-Devnagari and Devnagari-Malayalam has obtained 95 and 92 % identification rate correspondingly which is 0.7 and 3.7 % lower than the Eight-script average identification rate. The case of Bangla-Devnagari is due to some similar features in their writing style (presence of topological feature like 'matra' in both cases). Devnagari-Malayalam combination produces discouraging results due to

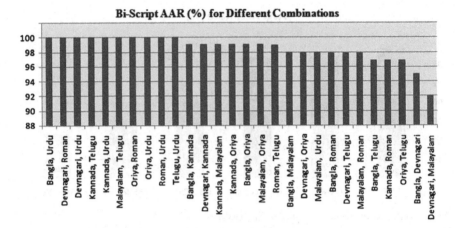

Fig. 8 Bi-script average accuracy rate (%) for different combinations

presence of few structural similarities of the graphemes of these two scripts. We need to investigate this issue in more detail and some fine set of features need to be developed for them. Hopefully we will achieve this very shortly.

Figure 8 shows a bar graph where the Bi-Script average accuracy rate of different Bi-Script combinations and their reported AAR (%) has been shown.

3.5 Comparative Study

To understand the performance of any proposed technique, we generally compare the standard performance measuring parameters namely average accuracy rate, model building time, etc. with other state-of-the-art available. But unfortunately we are unable to compare our results because no work is available till date for above mentioned handwritten Eight-script combination at Line level to the best of our knowledge. The issue of unavailability of benchmark dataset inspired us to prepare our own document image data bank. Our present result can be considered as a benchmark one for this Eight-script combination on the present Line-level dataset.

4 Conclusion and Future Scope

Different techniques have been proposed by the researchers for Indic script identification. But the solution to the handwritten script identification problem considering all official Indic scripts is still far from the complete solution. In this paper, an approach for automatic script identification from eight official Indic scripts has been proposed. The method is robust enough against standard skew and noise that presents in real-life handwritten documents. Experimental result shows an average

Eight-script identification rate of **95.7** % using MLP classifier which is really encouraging. Other classifiers have also shown comparable performance on our developed method. Experimentation on $^8C_2$ or 28 possible Bi-script combinations are also performed. Average Bi-script identification rate is found to be **98.51** %, which is very much promising in *HSI* category. The present result can be considered as a benchmark one for Eight-script combinations on the present dataset. The authors will be happy to contribute this dataset for the document image processing research community for non commercial use and will be available freely on request.

Script identification techniques may be applied on Page/Block/Line/Word/Character level. The overhead of segmentation is not applicable for Page level script identification. But in other cases namely Block/Line/Word/Character level, the accuracy rate solely depends on the performances of the segmentation algorithm followed. Researchers sometimes considered pre-segmented images for their experimentation. This is because segmentation itself is another broad area of research in document image processing. The problem becomes extremely challenging when handwritten documents are considered. The present work followed a semi-automatic segmentation technique while extracting lines from handwritten document images. Special care was taken to handle touching or skewed handwritten text lines. All the experimentation carried out is based on the pre-segmented output line-level images.

Availability of benchmark database is a real constraint in document image processing research. Though printed document may be available from some easy sources like news paper, book chapters, etc. but collection of handwritten samples is more challenging because of psychological barriers of the contributors who are writing to generate the corpus. So, future plan of the authors includes building benchmark dataset for all official handwritten Indic scripts. Scopes can be further extended to work on real-life script identification problem namely video-based script identification, script identification from scene images, character-level script identification from multi-script artistic words (few samples are shown in Figs. 9 and 10), etc.

Fig. 9 Character-level multi-script artistic document images

Fig. 10 Real life video script images

References

1. Obaidullah, S.M., Das, S.K., Roy, K.: A system for handwritten script identification from indian document. J. Pattern Recogn. Res. **8**(1), 1–12 (2013)
2. Ghosh, D., Dube, T., Shivprasad, S.P.: Script recognition—a review. IEEE Trans. Pattern Anal. Mach. Intell. **32**(12), 2142–2161 (2010)
3. Pal, U., Chaudhuri, B.B.: Identification of different script lines from multi-script documents. Image Vis. Comput. **20**(13-14), 945–954 (2002)
4. Hochberg, J., Kelly, P., Thomas, T., Kerns, L.: Automatic script identification from document images using cluster-based templates. IEEE Trans. Pattern Anal. Mach. Intell. **19**, 176–181 (1997)
5. Chaudhury, S., Harit, G, Madnani, S., Shet, R.B.: Identification of scripts of Indian languages by combining trainable classifiers. In: Proceedings of Indian Conference on Computer Vision, Graphics and Image Processing, 20–22 Dec 2000, Bangalore, India (2000)
6. Pal, U., Chaudhuri, B.B.: Script line separation from Indian multi-script documents. IETE J. Res. **49**, 3–11 (2003)
7. Pati, P.B., Ramakrishnan, A.G.: Word level multi-script identification. Pattern Recogn. Lett. **29**(9), 1218–1229 (2008)
8. Obaidullah, S.M., Mondal, A., Das, N., Roy, K.: Structural feature based approach for script identification from printed Indian document. In: Proceedings of International Conference on Signal Processing and Integrated Networks, pp. 120–124 (2014)
9. Hochberg, J., Bowers, K., Cannon, M., Kelly, P.: Script and language identification for handwritten document images. Int. J. Doc. Anal. Recogn. **2**(2/3), 45–52 (1999)
10. Zhou, L., Lu, Y., Tan, C.L.: Bangla/English script identification based on analysis of connected component profiles. In: Lecture Notes in Computer Science, vol. 3872/2006, 24354 (2006). doi:10.1007/11669487_22
11. Singhal, V., Navin, N., Ghosh, D.: Script-based classification of hand-written text document in a multilingual environment. In: Research Issues in Data Engineering, p. 47 (2003)

12. Roy, K., Banerjee, A., Pal, U.: A System for word-wise handwritten script identification for indian postal automation. In: Proceedings of IEEE India Annual Conference, pp. 266–271 (2004)
13. Hangarge, M., Dhandra, B.V.: Offline handwritten script identification in document images. Int. J. Comput. Appl. 4(6), 6–10 (2010)
14. Hangarge, M., Santosh, K.C., Pardeshi, R.: Directional discrete cosine transform for handwritten script identification. In: Proceedings of 12th International Conference on Document Analysis and Recognition, pp. 344–348 (2013)
15. Pardeshi, R., Chaudhury, B.B., Hangarge, M., Santosh, K.C.: Automatic handwritten Indian scripts identification. In: Proceedings of 14th International Conference on Frontiers in Handwriting Recognition, pp. 375–380 (2014)
16. Obaidullah, S.M., Mondal, A., Das, N., Roy, K.: Script identification from printed Indian document images and performance evaluation using different classifiers. Appl. Comput. Intell. Soft Comput. 2014(Article ID 896128), 12 (2014). doi:10.1155/2014/896128
17. Mandelbrot, B.B.: The Fractal Geometry of Nature. Freeman, New York (1982)
18. Bradski, G., Kaehler, A.: Learning OpenCV. O'Reilly Med., California (2008)
19. Shiv Naga Prasad, V., Domke, J.: Gabor filter visualization. Technical Report, University of Maryland (2005)
20. Aleai, A., Nagabhushan, P., Pal, U.: A benchmark kannada handwritten document dataset and its segmentation. In: Proceedings of International Conference on Document Analysis and Recognition, pp. 140–145 (2011)
21. Hall, M., Frank, E., Holmes, G., Pfahringer, B., Reutemann, P., Witten, I.H.: The WEKA data mining software: an update. SIGKDD Explor. 11, 10–18 (2009)
22. Obaidullah, S.M., Das, N., Roy, K.: Gabor filter based technique for offline indic script identification from handwritten document images. In: IEEE International Conference on Devices, Circuits and Communication (ICDCCom 2014), Ranchi, India, pp. 1–5. doi:10.1109/ICDCCom.2014.7024723

Writer Verification on Bangla Handwritten Characters

Chayan Halder, Sk. Md. Obaidullah, Jaya Paul and Kaushik Roy

Abstract Writer Identification/Verification being a biometric personal authentication technique can be extensively used for personal verification. Currently, it has gained a renewed interest in researchers due to the promising prospect in real life applications like forensic, security, access control, etc. In the proposed work, we have modified and evaluated the performance of different textural features for writer verification on unconstrained Bangla isolated characters. A collection of 500 documents of isolated Bangla characters from 100 writers consisting of total 35,500 Bangla characters (25,500 alphabets + 5000 Bangla numerals + 5000 Bangla vowel modifiers) are used in this respect. The combination of features yields better performance. The evaluation of results shows that our method is effective and can be applied on large database.

Keywords Writer verification · Bangla handwriting analysis · Mahalanobis distance · Textural features

C. Halder (✉) · K. Roy
Department of Computer Science, West Bengal State University,
Barasat, Kolkata 700126, West Bengal, India
e-mail: chayan.halderz@gmail.com

K. Roy
e-mail: kaushik.mrg@gmail.com

Sk.Md.Obaidullah
Department of Computer Science and Engineering, Aliah University,
Kolkata, West Bengal, India
e-mail: sk.obaidullah@gmail.com

J. Paul
Department of Information Technology, Government College of Leather Technology,
Kolkata 700098, West Bengal, India
e-mail: jayapl2005@gmail.com

© Springer India 2016
R. Chaki et al. (eds.), *Advanced Computing and Systems for Security*,
Advances in Intelligent Systems and Computing 396,
DOI 10.1007/978-81-322-2653-6_4

1 Introduction

The authentication of the persons based on biometric techniques is a challenging problem which has been an active area of research over the years. Handwriting is one of the primitively used biometric techniques to authenticate an individual. Every individual has some certain degree of stability in their handwriting which enables the handwriting analyser to verify the writer. Writer verification is the task of authenticating the writer of an unknown handwritten document. Mostly this is done by the experts of handwriting analysis manually by visual examination of the documents, but the reliability of those decisions is not conclusive. Automation of this task is less attempted than identification as verification requires a local decision-making techniques which is generally more dependent on the content of the writing. This writer verification system can be developed using text-dependent input data set or using text-independent input data set. In text-dependent writer verification system the system is dependent on a given text content. In a text-dependent method the known and unknown writers need to write the same text accordingly and the system matches the same characters and texts to verify the writer. The text-independent methods are able to identify writers independent of the text content and the system uses a generalized method to verify writers by finding similarities between writing patterns. Text-independent methods have got a wider applicability, but text-dependent methods have obtain higher accuracy. Text-dependent system can increase the possibility of forgery due to text dependency, but it can be very useful in case of low security applications or applications where genuine user uses the system most of the time where frequent rejection is not suitable. In case of text-independent method, possibility of forgery is less but it needs more input data set and rejection can be possible in case of genuine user also. The writer verification can be used in different fields like security, access control, graphology, historical document analysis [1] and hand held, and mobile devices [2]. To some extent it can be considered as strong as DNA and fingerprints in terms of verification [3].

The paper is outlined as follows: a brief overview of the significant existing contribution on writer identification/verification techniques is discussed in Sect. 2. A summarized description of the proposed method is presented in Sect. 3. In Sect. 4 data collection and preprocessing steps are described. The description of features that are used for the current work can be found in Sect. 5. Section 6 describes about verification methodologies followed by results in Sect. 7. At last conclusion is presented in Sect. 8.

2 Brief Survey on Writer Verification/Identification

Writer verification technique is quite similar to signature verification and very close to writer identification but there exists a certain dissimilarities between them which make it a different problem in document analysis domain. Various works can be found in the literature on automatic signature verification since 1989 [4–12]. There is a lot of progress in signature verification but most of them are based on online mode as seen in [6]. Working on verification techniques in offline mode is more challenging than in online mode. In online mode prior information of strokes and their starting and ending points are available but not in case of offline mode. Different systems can be found regarding writer identification systems [13–27]. Most of the works are on Roman script [13–16] like the works of Srihari et al. [13], Said et al. [14], Bulacu et al. [15] and Siddiqi and Vincent [16], etc. Jain and Doermann proposed multi script writer identification on English, Greek, and Arabic languages [17]. In [18], Ghiasi and Safabakhsh presented text-independent writer identification using codebook on English and Farsi handwriting database. Some writer identification works can be found in [19, 20]. Djeddi et al. in [21] have proposed writer identification on multiple languages like English and Greek using ICFHR 2012 Latin/Greek database. Recently, Halder et al. [22] have proposed writer identification on Devanagari script. They have used isolated characters for their work. Ding et al. in [23] have worked on Chinese, English, and Greek languages for writer identification. Also there are few writer identification works on Bangla scripts which can be seen in [24–27]. Garain and Paquet [24] proposed an AR coefficient feature-based writer identification system for Roman and Bangla script. Chanda et al. [25] have developed a text-independent writer identification system on Bangla script using the Gaussian kernel SVM (Support Vector Machine) as their classifier. They experiment their work on 104 writers and got 95.19 % accuracy on their system. Halder and Roy in [26] used only isolated Bangla numerals and in [27] used all characters (alphabets + numerals + vowel modifiers) for writer identification from 450 documents of 90 writers using 400 and 64 dimensional features and LIBLINEAR and MLP classifiers. In the work of [26] highest writer identification accuracy of 97.07 % has been achieved while in [27] 99.75 % accuracy has been achieved for the same. But the work on writer verification is very rare. Though there are few contributions on writer verification but most are in non-Indic scripts [28–35]. Yamazaki et al. [28] have proposed an online Writer verification process using hidden Markov models (HMM) on Chinese and Japanese characters. They have used 20 writers each with 20 different characters to generate the code book text. A single word-based writer identification and verification has been implemented by Zois and Anastassopoulos [29]. Experiments have been performed on a data set of 50 writers. An English word and its corresponding Greek word with same meaning and length were used. 20-dimensional feature vectors have been used for the work. Bayesian classifier and Multilayer Perceptron classifier have been used to test the efficiency of their approach. For Bayesian classifier they have got identification accuracy of 92.48 % on English and 92.63 %

on Greek words. For MLP the accuracies have been increased to 96.5 % and 97 % on English and Greek, respectively. They have achieved verification accuracies of 97.7 % and 98.6 %, respectively. As their approach is dependent on the dimensionality of the feature vector which is dependent on the length of the word so the success is also dependent on the word length, length of the SEs, and partition levels. As the results for both the languages are very close so it can be concluded that the approach can be language independent. In [30], Srihari et al. proposed different macro features with Gaussian and Gamma parameters along with log-likelihood ratio (LLR) on English handwritten samples from 1000 writers for writer verification. Bulacu et al. in [31] implemented writer identification and verification on Roman script using three different databases namely IAM database [32], Firemaker set [33], Unipen database [34]. The IAM database contains 650 writers, the Firemaker set contains 250 writers and the Unipen database contains 215 writers. They have also combined the IAM and Firemaker databases to form a database namely Large with 900 writers. The four different features directional probability distribution functions (PDFs), grapheme emission PDF, run-length PDFs, and autocorrelation are used by them. In another work by combining some textural and allographic features they have proposed text-independent Arabic writer identification [35]. The IFN/ENIT dataset [36] has been used for their work. For the allographic features, a codebook of 400 allographs has been generated from the handwritings of 61 writers and the similarities of these allographs have been used as another feature. The database has been collected from 350 writers with five samples per writer [each sample contains two lines (about 9 words)]. The best accuracies that have been seen in experiments are 88 % in top-1 and 99 % in top-10. There are very few works done on Indic script like Gupta and Namboodir [37] proposed a writer verification system using boosting method on Devanagari script on 20 writers to achieve error rates of 5 % for 7 words, 11 % when chosen randomly, and 22 % for primitive selection methods. According to our knowledge there no methods attempted on writer verification considering Bangla script which motivates us to work in this area. In this proposed method, some modification on fast Fourier transform (FFT), discrete cosine transform (DCT) and gray level co-occurrence matrix (GLCM) is done to extract textural features from superimposed characters images. Though the selected feature set is small considering the writer verification modality, satisfactory results are achieved.

3 Method

In this section, a brief description and outline of the proposed system has been presented. For the current work we have considered a database consisting of isolated characters collected from 100 writers. The inter writer (the variation between the handwriting samples of two different people) and intra writer (the variation within a person's own handwriting samples) variation can be seen in handwriting of different writers. This difference can be seen in Fig. 1, where in Fig. 1a three different isolated

Fig. 1 **a** Example of three
different isolated characters
from three different writers.
b Example of three
superimposed characters of
each writer from the same
three writers. **c** Example of
three superimposed characters
from ten different writers
including the three writer
from (**a**) and (**b**)

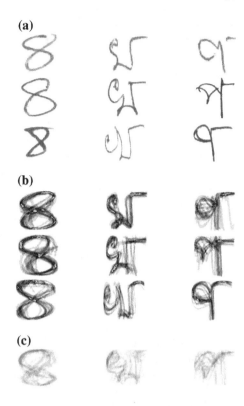

Fig. 1 **a** Example of three different isolated characters from three different writers. **b** Example of three superimposed characters of each writer from the same three writers. **c** Example of three superimposed characters from ten different writers including the three writer from (**a**) and (**b**)

characters from three different writers are shown. In Fig. 1b superimposed version of three characters of these three writes (writer wise) are shown and in Fig. 1c superimposed version of these three characters for 10 different writers (inter writer including the three writers) are shown. Here, the concentrated areas of characters of Fig. 1b are very high than in Fig. 1c due to the reason that the inter writer variation is very high compared to intra writer variation. This is due to the individuality property of the characters. The experts analyze and verify the writer on the basis of these visual differences. For more details on intra, inter writer variation, and individuality see [27]. Here we have used textural features to obtain that difference by means of automation for writer verification. After calculation of textural features, distance measure, and threshold are applied to verify an unknown writer from a set of known writers. Figure 2 gives a brief overview of the proposed writer verification technique. First, the collected data are preprocessed and superimposed in each character category and then textural features MFFT (modified fast fourier transform), MGLCM (modified gray level co-occurrence matrix), and MDCT (modified discrete cosine transform) are applied individually and combined.

Fig. 2 Brief overview of the
writer verification system

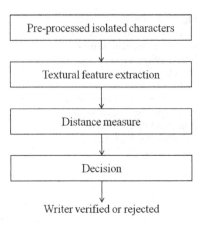

4   **Database and Preprocessing**

The current experiment is conducted using the database taken from [27] but with
more number of writers. The database contains 500 documents of isolated Bangla
characters from 100 writers consisting of total 35,500 Bangla characters (25,500
alphabets + 5000 Bangla numerals + 5000 Bangla vowel modifiers). An example of
our designed sample data collection form for isolated characters is shown in Fig. 3.
There exists no restrictions for writers regarding the type of pen and ink they have
used; some of them have used pencils also. The documents were scanned using a
flat-bed scanner and digitized in gray-scale mode at 300 dpi in TIF format. In
Preprocessing stage the global binarization of the whole document is carried out.
Now, maximum run length in both horizontal and vertical directions are calculated
to identify boundary lines in each directions and then these lines are being deleted.
After that, using bounding boxes and location information of the suggestive
characters, the isolated characters are extracted from the raw document images. The
details about the data collection, type of data, digitization of the raw collected
handwritten data and preprocessing techniques can be found in [27].

In our proposed work the isolated images are not directly used for feature
extraction. The characters of each writers are superimposed onto each other to
create a single character image of each character category belong to that writer only.
It means the same characters of a writer from five different sets are taken and
superimposed to create a single character of that writer which contains the intra
writer variation of that writer for that particular character. First, the bounding box of
isolated gray character images are calculated. After applying global binarization the
images are normalized to fixed 128 × 128 pixels size. Next, the normalized
128 × 128 binary images are projected into a white 128 × 128 image. In the
projection technique, for each object pixel of the original image, corresponding
pixel of the white 128 × 128 image has been decremented by a fixed value that is
calculated using the formula (1). Thus, a single gray character image is created
capturing the writing variation. This procedure is repeated for each character

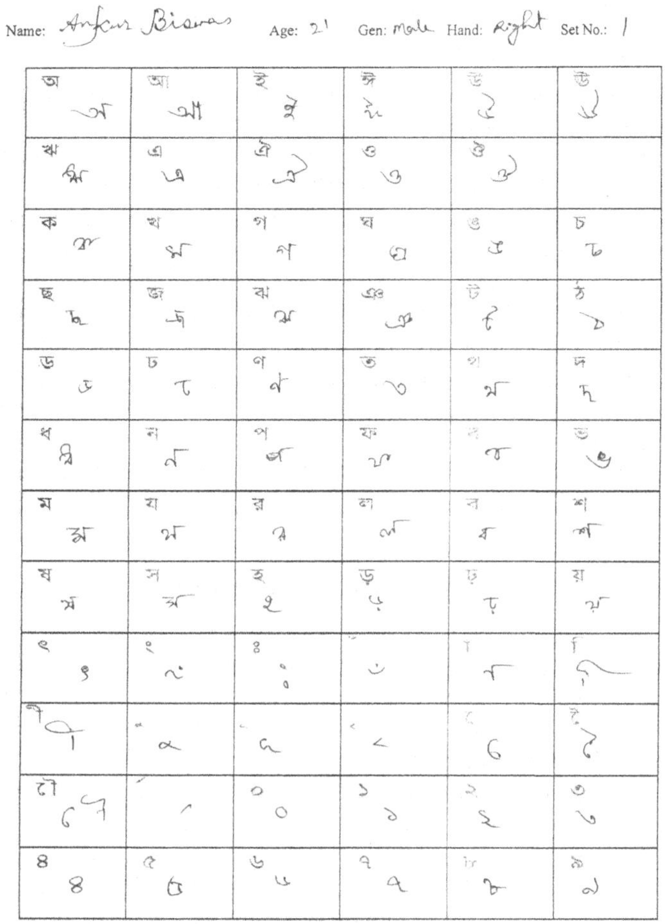

Name: Anfar Biswas   Age: 21   Gen: male   Hand: Right   Set No.: 1

Fig. 3 Sample data collection form used for collection of Bangla Handwritten isolated characters and Vowel modifiers

category writer wise. Figure 1b shows some sample superimposed images that are used for feature extraction.

$$\left\lceil \frac{N}{s} \right\rceil \tag{1}$$

where N Total number of gray levels

where s Total number of image samples that are used to create a superimposed image.

5 Feature Extraction

In this proposed work, the textural features of the superimposed images are extracted to distinguish between the intra writer and inter writer variations. Two frequency domain features FFT (fast Fourier transform) and DCT (discrete cosine transform) are modified along with GLCM (gray level co-occurrence matrix) to extract the textural features. In general, the FFT and DCT are used to capture the varying frequency of an image but in this experiment the varying gray level intensity (gray-level frequency) values, i.e., the textural differences are calculated by modifying FFT, DCT, and GLCM. The MFFT (modified fast Fourier transform) is used to get the variation of gray level intensity values. The MGLCM (modified gray level co-occurrence matrix) is used to get the local variation among gray level pixel values, probability of occurrence, uniformity, closeness of the distribution of the gray-level pixel values. The MDCT (modified discrete cosine transform) is used to get the similar textural measures like MFFT but with less computational cost.

5.1 MFFT (Modified Fast Fourier Transform)

The Fourier transform has many different variations. Among those the discrete Fourier transform (DFT) is very widely used. The fast Fourier transform (FFT) is a quicker version of DFT where the computational overhead is lower compared to DFT. Using FFT pixel values of an image along a row or column can be transformed into a periodic sequence of complex number. To obtain image information of the frequency domain space FFT can be used. After a Fourier decomposition of the image, the phase spectrum contains texture and structure information about the image. More details about FFT can be found in [38]. The 2D FFT function computes transformation of a given 2D image of length $M \times N$ using the Eq. (2).

$$F(x,y) = \sum_{m=0}^{M-1} \sum_{n=0}^{N-1} f(m,n) e^{-j2\pi(x\frac{m}{M} + y\frac{n}{N})} \tag{2}$$

In the current method the Modified FFT feature is calculated on the superimposed images using the following steps: First, the 128×128 dimensional feature vectors has been computed using 2D FFT algorithm then using Gaussian filter and Eq. (3) MFFT is calculated on the images to get 64 dimension feature vectors.

$$F(x) = \left\{ \frac{f(m)}{M}, F(x) \le 1 \right\} \tag{3}$$

where

$$M = \max(f(m))$$

$$f(m) = \sum_{n=1}^{N} m_n$$

and

N Total number of feature dimension for a single column of the feature set.

5.2 MGLCM (Modified Gray Level Co-occurrence Matrix)

The GLCM (gray level co-occurrence matrix) is a statistical calculation of how often different combination of gray level pixel values occur in an image. It has been the workhorse for textural analysis of images since the inception of the technique by Haralick et al. [39]. GLCM matrix describes the frequency of occurrence of one gray level with another gray level in a linear relationship within a defined area. Here, the co-occurrence matrix is computed based on two parameters, which are the relative distance between the pixel pair d measured in pixel number and their relative orientation φ. Normally, φ is quantized in four directions (0°, 45°, 90°, and 135°). The GLCM is a matrix where the number of rows and columns are equivalent to the number of gray levels of the image. The matrix element $P(i, j|\Delta x, \Delta y)$ is the relative frequency with which two pixels, separated by a pixel distance (Δx, Δy), occur within a given neighborhood, one with intensity i and the other with intensity j. One may also say that the matrix element $P(i, j|d, \theta)$ contains the second-order statistical probability values for changes between gray levels i and j at a particular displacement distance d and at a particular angle (θ). Detail description of GLCM may be available in [39]. In the current approach the MGLCM is calculated with

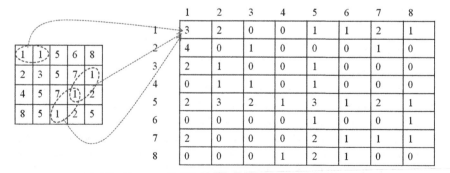

	1	2	3	4	5	6	7	8
1	3	2	0	0	1	1	2	1
2	4	0	1	0	0	0	1	0
3	2	1	0	0	1	0	0	0
4	0	1	1	0	1	0	0	0
5	2	3	2	1	3	1	2	1
6	0	0	0	0	1	0	0	1
7	2	0	0	0	2	1	1	1
8	0	0	0	1	2	1	0	0

Fig. 4 GLCM calculation for both type of pixel pairs in all four directions

contrast, correlation, energy, and homogeneity statistical measures in all four directions considering both type of pairs like $P[i, j]$ and $P[j, i]$. The following Fig. 4 is showing the GLCM calculation technique in this respect for all four directions and both pixel pair types considering eight gray levels. After the calculation of GLCM using the Eq. (3) the MGLCM feature vectors are calculated to get eight dimensional features.

5.3 MDCT (Modified Discrete Cosine Transform)

Discrete cosine transform (DCT) is one of the widely used transform in the image processing applications for feature extraction. The approach involves taking the transformation of the image as a whole and separating the relevant coefficients. DCT performs energy compaction. The DCT of an image basically consists of three frequency components namely low, middle, high each containing some detail information in an image. The low frequency generally contains the average intensity of an image. The DCT is very similar to FFT. The main difference between a DCT and a DFT is that the DCT uses only cosine functions, while the DFT uses both sin and cosine. To get more details on DCT see [40]. The 2D DCT function computes the transformation of a given $M \times N$ image by using the Eq. (4). The 64 dimensional MDCT feature vectors of the superimposed images has been calculated using the following steps: first, the 128×128 dimensional feature vector has been computed using 2D DCT then Gaussian filter and Eq. (3) is used.

$$F(x, y) = (\frac{2}{M})^{\frac{1}{2}}(\frac{2}{N})^{\frac{1}{2}} \sum_{m=0}^{M-1} \sum_{n=0}^{N-1} \Lambda(m)\Lambda(n)$$
$$\cos[\frac{\pi x}{2M}(2m+1)] \cos[\frac{\pi y}{2N}(2n+1)]f(m,n) \tag{4}$$

where
$f(m, n)$ The gray intensity value of the pixel (m, n).

6 Verification

In this study, the writer verification is performed using Mahalanobis distance measure technique between features of known writers of the database and the unknown writer. In the proposed system, the textural features are extracted from the isolated handwritten characters of the unknown writer and after calculating Mahalanobis distance the varying decision threshold is applied to verify the writer. During the verification phase the features extracted from the known writers are used to calculate the distance between the known and questioned writer. The feature vectors of the known and questioned writer in the feature space during a certain comparison can be defined as follows:

$$\overleftarrow{KW_j} = (kf_{1j}, kf_{2j}, kf_{3j}, \ldots, kf_{Nj}) \tag{5}$$

and

$$\overleftarrow{QW} = (qf_1, qf_2, qf_3, \ldots, qf_N)$$

where

kf_{ij} and qf_i represent each features of the known writer and the questioned writer, respectively.

and

N is the feature dimension.

Mahalanobis distance on each of these feature vector pair has been calculated which can be defined as follows:

$$D(x) = \sqrt{(kf_{ij} - qf_i)^T S^{-1} (kf_{ij} - qf_i)} \tag{6}$$

where kf_{ij} and qf_i are same as defined in (5)

and

S^{-1} is the inverse covariance matrix of the feature data set.

The decision threshold has been calculated after the distance measure and if the distance is greater than the threshold then the questioned writer is verified.

7   Results

The proposed experiment has been carried out on total 35,500 Bangla characters from 100 writers taking five samples of isolated characters from each writer. The textural features MFFT, MGLCM, and MDCT are used for feature extraction and Mahalanobis distance measure and varying decision threshold has been used to verify the writers. In the current system from 100 writers four samples are used to create the training feature set and the remaining samples are selected at random as the set of unknown writers. The features capture different aspects of handwriting individuality. By combining all the features improvement in the performance of verification has been achieved.

7.1   Result of Writer Verification

In order to get a reliable result two types of errors are considered along with two types of accuracy calculations. The true acceptance rate (TAR), (where the questioned writer is properly accepted) and the true rejection rate (TRR), (where the

Table 1 Writer verification result on different textural features

Features	Dimension	Accuracy (%)	Error (%)	
			FAR	FRR
MFFT	64	56.27	31.75	11.98
MGLCM	8	26.37	46.68	26.95
MDCT	64	48.57	35.33	16.09
MFFT + MGLCM	72	57.84	28.59	13.57
MFFT + MDCT	128	60.25	27.17	12.58
MGLCM + MDCT	72	52.42	29.00	18.58
MFFT + MGLCM + MDCT	136	62.17	26.22	11.61

questioned writer is properly rejected) are summed up to get the accuracy value. The false acceptance rate (FAR), (where the questioned writer is not the original writer and the system wrongly accepted it) and false rejection rate (FRR), (where the questioned writer is the original writer but the system wrongly rejected it) are considered as the total error rate. Table 1 shows the writer verification results in terms of individual features and combination of features. From the table it can be observed that in case of individual features MFFT scores highest with 56.27 % verification accuracy where GLCM scored very low due to the very high False acceptance rate (FAR) of 46.68 %. When the three features are combined the accuracy has been increased to 62.17 % which is quite satisfactory.

7.2 Comparative Study

In the current method, we have used 100 writers and textural features. According to our knowledge of the literature there is no such work of writer verification on Bangla script so currently we are unable to compare our experiment with others. In our current approach three different textural features are applied. A comparative study on writer verification results of these features and their combinations can be seen in Fig. 5. In Fig. 6 a comparative analysis of different features and their combination with respect to accuracy and error rate is shown. Analyzing these figures, it can be seen that the MFFT feature gives better performance compared to others regarding single and combined use. The performance of MGLCM feature is quite low due to the higher False Acceptance Rate (FAR), while performance of MDCT is moderate. When combination of different features is used the accuracy is improved and error rate are also decreased. Though the results are not very high due

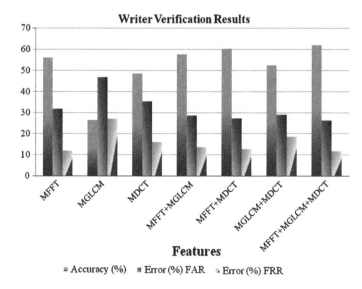

Fig. 5 Comparison of accuracy and error rate of writer verification results with respect to different features and their combinations

to high false acceptance rate (FAR) but the results show some encouragement toward the works of handwritten Bangla writer verification. Our approach is open for comparison with other works.

8 Conclusion

In this document a study has been conducted on Bangla isolated characters for writer verification. The lack of standard Bangla handwriting analysis system and a standard Bangla database with writer information, has initiated our interest in this work. In the current work, we have used a database consists of 100 writers and modified some simple frequency domain features to use as textural features for our approach. Although the used feature set is small regarding writer verification modality, encouraging results are successfully achieved. At present there is no such writer verification system on Bangla with which we can compare our system, still it can be used as a stepping stone towards this type of works on Bangla. The amount of data is relatively large enough which indicates that this method can be applied in real-life environment.

Future scope includes increasing the size of the database both in terms of writers and samples per writer to create a standard database for the community of handwriting analysers. Also we are looking forward to introduce handwriting of skilled forgers to give an extra dimension to our work. In future fuzzy measures can be introduced during feature selection so that false acceptance rate can be minimized.

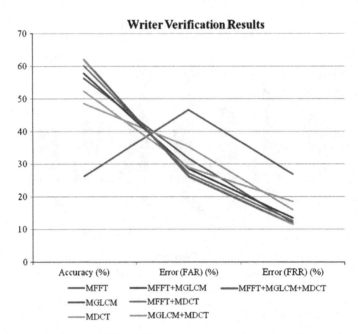

Fig. 6 Comparative study of different features and their combinations with respect to accuracy and error rate of writer verification results

We are also planning to use different kind of statistical distance measure and analysis methods for decision making during verification. We believe that this kind of approach can be applied on other similar Brahmic scripts.

Acknowledgments One of the author would like to thank Department of Science and Technology (DST) for support in the form of INSPIRE fellowship.

References

1. Fornes, A., Llados, J., Sanchez, G., Bunke, H.: Writer identification in old handwritten music scores. In: Proceedings of 8th IAPR workshop on DAS, pp. 347–353 (2008)
2. Chaudhry, R., Pant, S.K.: Identification of authorship using lateral palm print: a new concept. J. Forensic Sci. **141**, 49–57 (2004)
3. Schomaker, L.: Advances in writer identification and verification. In: Proceedings of 9th ICDAR, vol. 2, pp. 1268–1273 (2007)
4. Plamondon, R., Lorette, G.: Automatic signature verification and writer identification: the state of the art. Pattern Recogn. **22**(2), 107–131 (1989)
5. Chan, L.F., Kamins, D., Zimerann, K.: Signature recognition through spectral analysis. Pattern Recogn. **22**, 39–44 (1989)
6. Wan, L., Wan, B., Lin, Z.-C.: On-line signature verification with two stage statistical models. In: Proceedings of 8th ICDAR, pp. 282–286 (2005)

7. Prakash, H.N., Guru, D.S.: Offline signature verification: an approach based on score level fusion. Int. J. Comput. Appl. **1**(18), 52–58 (2010)
8. Yilmaz, M.B., Yanikoglu, B., Tirkaz, C., Kholmatov, A.: Offline signature verification using classifier combination of HOG and LBP features. In: Proceedings of IJCB, pp. 1–7 (2011)
9. Malik, M.I., Liwicki, M., Dengel, A.: Evaluation of local and global features for offline signature verification. In: Proceedings of 1st AFHA, pp. 26–30 (2011)
10. Liwicki, M., Malik, M.I., van den Heuvel, C.E., Xiaohong, C., Berger, C., Stoel, R., Blumenstein, M., Found, B.: Signature verification competition for online and offline skilled forgeries (SigComp2011). In: Proceedings of 11th ICDAR, pp. 1480–1484 (2011)
11. Ferrer, M., Vargas, J., Morales, A., Ordonez, A.: Robustness of offline signature verification based on gray level features. IEEE Trans. Inf. Forensics Secur. **7**(3), 966–977 (2012)
12. Neamah, K., Mohamad, D., Saba, T., Rehman, A.: Discriminative features mining for offline handwritten signature verification. 3D Res. **5**(1), 1–6 (2014)
13. Srihari, S., Cha, S., Arora, H., Lee, S.: Individuality of handwriting. J. Forensic Sci. **47**(4), 1–17 (2002)
14. Said, H.E.S., Tan, T.N., Baker, K.D.: Personal identification based on handwritting. Pattern Recogn. **33**, 149–160 (2000)
15. Bulacu, M., Schomaker, L., Vuurpijl, L.: Writer identification using edge-based directional features. In: Proceedings of 7th ICDAR, vol. 2, pp. 937–941 (2003)
16. Siddiqi, I., Vincent, N.: Writer identification in handwritten documents. In: Proceedings of 9th ICDAR, pp. 108–112 (2007)
17. Jain, R., Doermann, D.: Writer identification using an alphabet of contour gradient descriptors. In: Proceedings of 12th ICDAR, pp. 550–554 (2013)
18. Ghiasi, G., Safabakhsh, R.: Offline text-independent writer identification using codebook and efficient code extraction methods. Image Vis. Comput. **31**, 379–391 (2013)
19. Djeddi, C., Souici-Meslati, L., Ennaji, A.: Writer recognition on Arabic handwritten documents. In: Proceedings of 5th ICISP, pp. 493–501 (2012)
20. Abdi, M.N., Khemakhem, M.: A model-based approach to offline text-independent Arabic writer identification and verification. Pattern Recogn. **48**(5), 1890–1903 (2015)
21. Djeddi, C., Siddiqi, I., Souici-Meslati, L., Ennaji, A.: Text-independent writer recognition using multi-script handwritten texts. Pattern Recogn. Lett. **34**, 1196–1202 (2013)
22. Halder, C., Thakur, K., Phadikar, S., Roy, K.: Writer Identification from handwritten Devanagari script. In: Proceedings of INDIA-2015, pp. 497–505 (2015)
23. Ding, H., Wu, H., Zhang, X., Chen, J.P.: Writer identification based on local contour distribution feature. Int. J. Signal Process. Image Process Pattern Recogn. **7**(1), 169–180 (2014)
24. Garain, U., Paquet, T.: Off-line multi-script writer identification using AR coefficients. In: Proceedings of 10th ICDAR, pp. 991–995 (2009)
25. Chanda, S., Franke, K., Pal, U., Wakabayashi, T.: Text independent writer identification for Bengali script. In: Proceedings of ICPR, pp. 2005–2008 (2010)
26. Halder, C., Roy, K.: Individuality of isolated Bangla numerals. J. Netw. Innov. Comput. **1**, 33–42 (2013)
27. Halder, C., Roy, K.: Individuality of isolated Bangla characters. In: Proceedings of ICDCCom, pp. 1–6 (2014)
28. Yamazaki, Y., Nagao, T., Komatsu, N.: Text-indicated writer verification using hidden Markov models. In: Proceedings of 7th ICDAR, pp. 329–332 (2003)
29. Zois, E., Anastassopoulos, V.: Morphological waveform coding for writer identification. IEEE Trans. Pattern Recogn. **33**(3), 385–398 (2000)
30. Srihari, S.N., Beal, M.J., Bandi, K., Shah, V.: A statistical model for writer verification. In: Proceedings of 8th ICDAR, pp. 1105–1109 (2005)
31. Bulacu, M., Schomaker, L.: Text-independent writer identification and verification using textural and allographic features. IEEE Trans. Pattern Anal. Mach. Intell. **29**(4), 701–717 (2007)

32. Marti, U., Bunke, H.: The IAM-database: an english sentence database for offline handwriting recognition. Int. J. Doc. Anal. Recogn. **5**(1), 39–46 (2002)
33. Schomaker, L., Vuurpijl, L.: Forensic writer identification: a benchmark data set and a comparison of two systems. NICI, Nijmegen (2000). technical report
34. Guyon, I., Schomaker, L., Plamondon, R., Liberman, R., Janet, S.: UNIPEN project of online data exchange and recognizer benchmarks. In: Proceedings of 12th ICPR, pp. 29–33 (1994)
35. Bulacu, M., Schomaker, L., Brink, A.: Text-independent writer identification and verification on offline arabic handwriting. In: Proceedings of 9th ICDAR, pp. 769–773 (2007)
36. Pechwitz, M., Maddouri, S., Margner, V., Ellouze, N., Amiri, H.: IFN/ENIT-database of handwritten arabic words. In: Proceedings of CIFED, pp. 129–136 (2002)
37. Gupta, S., Namboodiri, A.: Text dependent writer verification using boosting. In: Proceedings of 11th ICFHR (2008)
38. Duhamel, P., Vetterli, M.: Fast Fourier transforms: a tutorial review and a state of the art. Sig. Process. **1**(9), 259–299 (1990)
39. Haralick, R.M., Shanmugan, K., Dinstein, I.: Textural features for image classification. IEEE Trans. Syst. Man Cybern. **SMC-3**(6), 610–621 (1973)
40. Ahmed, N., Natarajan, T., Rao, K.R.: Discrete cosine transform. IEEE Trans. Comput. **C-23**(1), 90–93 (1974)

Face Recognition in Video Using Deformable Parts Model with Scale Invariant Feature Transform (DPSIFT)

V. Mohanraj, V. Vaidehi, Ranajith Kumar and R. Nakkeeran

Abstract Face recognition is a complex task due to the challenges of varying pose, illumination, scaling, rotation, and occlusion in live video feed. This paper proposes a hybrid approach for face recognition in video called Deformable Parts Model with Scale Invariant Feature Transform (DPSIFT), to make face recognition system invariant to illumination, scaling, rotation, and limited pose. The proposed method identifies the significant points of the face using deformable part model and SIFT feature descriptors are extracted for those significant points. Fast Approximate Nearest Neighbor (FLANN) algorithm is used to match the SIFT descriptors between gallery image and probe image to recognize the face. The proposed method is tested with video datasets like YouTube celebrities, FJU, and MIT-India. DPSIFT method was found to perform better than the existing methods.

Keywords Face detection and recognition · SIFT · Deformable parts model · FLANN

V. Mohanraj (✉) · V. Vaidehi
Madras Institute of Technology, Anna University, Chennai, India
e-mail: mohanraj4072@gmail.com

V. Vaidehi
e-mail: vaidehi@mitindia.edu

R. Kumar
BARC, Mumbai, India
e-mail: ranajitk@barc.gov.in

R. Nakkeeran
Christ College of Engineering and Technology, Pondicherry, India
e-mail: sudhandhiram64@gmail.com

© Springer India 2016
R. Chaki et al. (eds.), *Advanced Computing and Systems for Security*,
Advances in Intelligent Systems and Computing 396,
DOI 10.1007/978-81-322-2653-6_5

1   Introduction

Face recognition (FR) has received significant attention during the past years. This is due to its use in a wide range of commercial and law enforcement applications. Computational complexity is the major bottleneck for real-time face recognition in video. Also, a face recognition system in real-time environment should handle the differences in the training and test database.

Face recognition system is a biometric software application for automatically identifying or verifying a person from a digital image by comparing with a stored database of faces. Face recognition includes both face identification and face verification, (i.e., authentication). Face verification is concerned with validating a claimed identity based on the image of a face, and either accepting or rejecting the identity claim. It computes similarity with only the claimed class in the database. This approach is one-to-one matching. Face identification is identifying a person based on the image of a face. This face image has to be compared with all the registered persons, thus increasing the computation time. This approach is one-to-many matching. There are two phases in face recognition, namely training and testing phase. In training phase, all the registered and authenticated persons in the database are processed to obtain facial features. These facial features are stored in the database for testing purpose. In testing phase, images of different persons are processed in the same manner as in the training phase and feature descriptors are compared with the database features. Similarity measures are used to verify or identify the test face with the database face.

There are basically two types of approaches in face recognition, the holistic approach and the feature-based approach. Holistic methods use the whole face region as the raw input to a recognition system. In feature-based methods, compressed features (geometric and/or appearance) are first extracted and fed into a structural classifier. Just as the human perception system uses both local features and the whole face region to recognize a face, a robust hybrid face recognition system takes the best advantage of both the features. Face Recognition System (FRS) has three main steps: face detection from input image or video, face recognition by comparison with database, and authorization results using similarity measures.

This paper proposes a novel approach for face recognition called DPSIFT. Deformable part model is used to localize the prominent features in face and scale invariant feature transform is used to extract the feature descriptor from interest point. Fast approximate nearest neighbor is used to match the feature descriptor between gallery image and probe image. This paper is organized as follows: Sect. 2 deals with related works, Sect. 3 elucidates the proposed DPSIFT method, and Sect. 4 presents the details of implementation and results. Finally, Sect. 5 concludes the paper.

2 Related Works

Face detection and recognition in video is a challenging task due to large variations in the face compared to gallery face images. In the past few decades many works on face recognition have been developed, yet the accuracy was not good due to large variations in illumination, scale, rotation, and occlusion so the system failed to recognize the person. Face recognition is classified into holistic features, local features, and hybrid features. Recently most works on face recognition using prominent features is used to identify the person. The prominent features will reduce the feature dimension. This paper also uses the prominent features for face identification, challenges in identification of prominent features is difficult due to variation in pose, illumination, scale, and occlusion.

The intersection of two edges [1] for two distinct directions are represented as interest points in Harris corner detection. The Harris corner method is invariant to rotation of the image. If the image is rotated then the corners are detected with the same interest points, but is variant to scale. Lowe et al. [2] suggests Scale Invariant Feature Transform (SIFT) to identify the interest points. The interest points are localized by constructing three levels of pyramids, i.e., below one scale and above one scale input image size. Interest points are invariant to scale, rotation, noises, and illumination changes. However, it consumes more time for interest point detection. In order to reduce time for interest points detection [3] suggested a method called Speeded up Robust Features (SURF). It is invariant to rotation and scale, but SIFT and SURF are not open source for commercial purpose.

A joint statistical method is suggested to estimate a shape of face image [5]. This method detects eyes, nose, and mouth part of face image by mapping the given input image and estimated shape of face image. It provides a large set of features for face image. However, it requires high resolution images and large training annotation dataset at high cost. The error between the estimated shape of face image with the given face image fails to predict the interest point's localization under tilted face image conditions. Another method [4] for interest point's localization is Viola-Jones algorithm based on the adaboost classifier. The method consists of a set of training images for eyes, nose, and mouth features of face image. It locates the interest points independently using adaboost classifier for each of the features by trained dataset for the input face image.

The Deformable Parts Model (DPM) goes one step further by fusing the local appearance model and the geometrical constraint into a single model. The DPM is given by a set of parts along with a set of connections between certain pairs of parts arranged in a deformable configuration. A natural way to describe the DPM is an undirected graph with vertices corresponding to the parts and edges representing the pairs of connected parts. The DPM-based detector estimates all landmark positions simultaneously by optimizing a single cost function composed of a local appearance model and a deformation cost. The complexity of finding the best landmark configuration depends on the structure of the underlying graph. If the graph does not contain loops, e.g., it has star-like structure with the central node corresponding to the nose, the estimation can be solved efficiently by a variant of dynamic programming. The proposed system combines the deformable parts model and scale invariant feature transform for face recognition in video.

3 Proposed System

The proposed system combines deformable parts model and scale invariant feature transform for efficient face recognition in video. The proposed method has two phases, namely training and testing. Figure 1 shows the framework of the proposed method.

3.1 Face Detection

Face detection is the first stage in face recognition system. The proposed DPSIFT method uses haar features for face detection in video as it works faster due to rejection of non-face images at the earliest stage and thus is suited for face detection in real time. Figure 2 shows some of the haar features used for face detection.

3.2 Interest Point Detection

The proposed DPSIFT method uses the deformable parts model to find the prominent feature points in the detected face image. The center point of the detected face is identified by referring to the bounded rectangle of the detected face. Further, the keypoints are localized in the extracted face image by matching a ground truth template upon the identified center point. Figure 3 shows the ground truth model for localizing the prominent feature points.

The center point lies upon the nose tip of detected face image and it is denoted as 'S_0' as shown in Fig. 3. The left corner of the left eye is represented as S_5, the right corner of the left eye is represented as S_1, the left corner of the right eye is represented as S_2, and the right corner of the right eye is represented as S_6. The left and right corners of the mouth region are represented as S_3 and S_4.

Let $J = \{I_1, I_2, ..., I_n\}$ be a set of grayscale images, where $\{I_i\}$ is an image of size $\{h \times w\}$ pixels, where 'h' denotes height and 'w' denotes width of an image. The localized set of N prominent feature is represented as a graph where $N = G(V, E)$, where $V = \{0,. .. ., N-1\}$ is a set of prominent feature points. The quality of prominent feature points is measured by Eq. 1. Figure 4 shows prominent feature points localization after face detection.

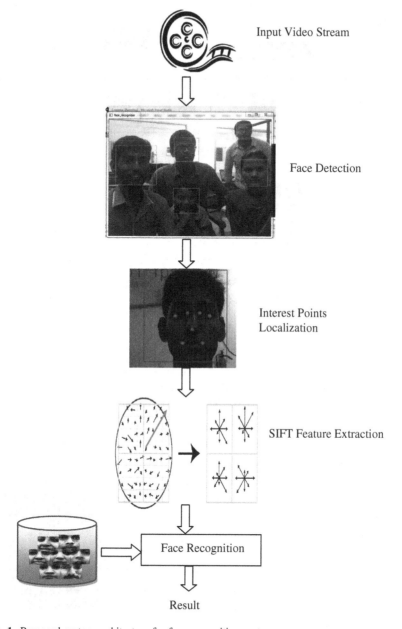

Input Video Stream

Face Detection

Interest Points
Localization

SIFT Feature Extraction

Face Recognition

Result

Fig. 1 Proposed system architecture for face recognition system

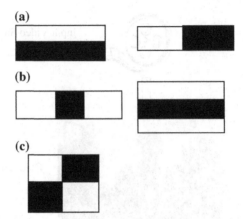

Fig. 2 Haar features (a) Edge Features, (b) line Features, (c) four rectangle features

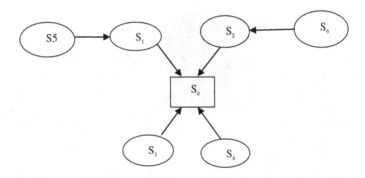

Fig. 3 List of prominent feature points

Fig. 4 Localized facial
feature points

$$f(I, S) = \sum_{i \in V} q_i(I, S_i) + \sum_{(i,j) \in E} g_{ij}(S_i, S_j) \qquad (1)$$

The first part of this equation identifies the prominent feature point for the
detected face image by ground truth. The second part of this equation evaluates the

Fig. 5 SIFT feature
descriptor

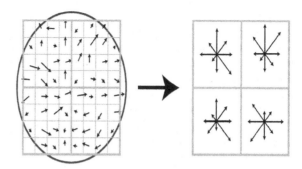

deformation cost by considering neighborhood pixel value for each prominent feature to locate the prominent feature points in the proper location.

3.3 Feature Descriptors

The proposed DPSIFT method uses scale invariant feature transform (SIFT) to extract the feature descriptor for each prominent feature point. After detection of prominent feature points from the cropped face image, the SIFT feature descriptor is applied to extract the features. A 16×16 neighborhood is constructed around each prominent feature point and it is divided into 16 sub blocks of 4×4 size. For each sub block 8 bin orientation histogram is calculated, so a total of 128 bins histogram are created for each prominent feature point. Figure 5 shows the feature descriptor for each prominent feature point.

3.4 Descriptors Matching

The proposed DPSIFT method uses fast approximate nearest neighbor (FLANN) to match the feature descriptor between gallery image and probe image. FLANN algorithm works faster than brute force matcher algorithm for larger database. The descriptor of one feature is taken in probe set and matched with all other features in the gallery set using Euclidian distance and the closest one is identified as matching descriptor.

4 Implementation and Results

The proposed DPSIFT method is implemented in Alienware Intel core i7 processor, 6 GB RAM, and 1 TB hard disk with Open Source Computer Vision library (OpenCV). The proposed system contains two phases, training and testing phase.

3000 positive and 5000 negative images were collected for face detection training. Positive images mean images that contains face and negative means other than face. Viola–Jones algorithm is used for face detection in video. The proposed DPSIFT method is tested with YouTube celebrity, FJU and MIT—India Video face dataset. Figures 6, 7, and 8 show the sample video face dataset taken for testing of face recognition system. MIT–India face dataset is created by author to test the real-time video-based face recognition system. The dataset consists of a variety of poses, different illumination conditions, and scaling. The range for the face to be detected should be between 30 cm and 3.5 m approximately and illumination should be a minimum of 50–70lux.

Figures 9 and 10 show Viola–Jones face detection algorithm for MIT-India and FJU video face dataset.

After the detection of face, deformable parts model is applied to identify the prominent feature points in face image. Figures 11 and 12 show the prominent feature points localization for MIT-India and FJU video face dataset.

After the detection of prominent feature points in face image, SIFT feature descriptor is applied to extract the feature for face recognition. Figure 13 shows existing SIFT interest point detection and feature descriptors matching.

During testing phase the probe image feature descriptors is compared with gallery image feature descriptors using fast approximate nearest neighbor (FLANN). Figure 14 shows proposed DPSIFT method for feature descriptors

Fig. 6 YouTube celebrity face dataset

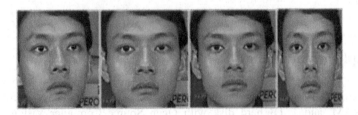

Fig. 7 FJU video face dataset

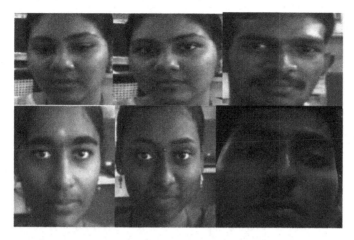

Fig. 8 MIT–India face dataset

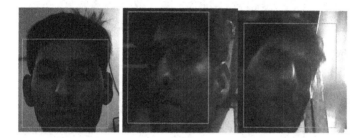

Fig. 9 Face detection for MIT–India face dataset

Fig. 10 Face detection for FJU face dataset

matching. Time taken for the interest point detection and feature descriptors matching process is less than the existing method.

DPSIFT algorithm is proposed to reduce the time required for interest points detection and feature descriptors matching between probe face image and gallery face image. DPSFIT algorithm is proposed in order to increase the performance of face recogntion with limited pose, invariant to illumination, orientation and scale.

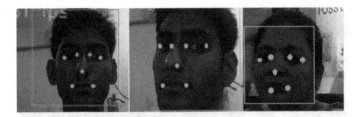

Fig. 11 Prominent feature points for MIT—India video face dataset

Fig. 12 Interest point detection for FJU video face dataset

Fig. 13 Existing SIFT interest points detection and descriptor matching

Fig. 14 Proposed interest points detection and feature descriptors matching

Fig. 15 Time complexity
analysis of DPSIFT method

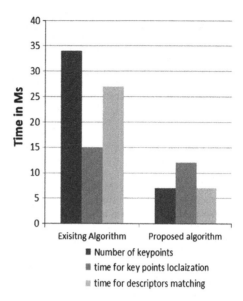

Number of keypoints
time for key points loclaization
time for descriptors matching

Table 1 Accuracy rate of
recognition for various
datasets

	Accuracy in %	
	Gabor	DPSIFT
MIT dataset 1	73	88.06
MIT dataset 2	70	84.77
FJU dataset	92.47	97.72
YT_video dataset	71.91	85.9

Figure 15 shows the performance of the DPSIFT method for FJU, YouTube
celebrity and MIT-India video face datasets. From the performance analysis, it is
observed that the number of key points is reduced by considering only the promient
features. These interest points are used to identify the person. The proposed
DPSIFT algorithm performs well for face recognition with reduced number of
features and takes lesser time to recognize the face. Figure 15 shows recognition
accuracy of the proposed DPSFIT method.

Table 1 shows that recognition rate of the proposed method gives better accuracy
than the existing method with three different datasets.

5 Conclusion

This paper proposes a hybrid approach for face recognition in video using
Deformable Parts Model with Scale Invariant Feature Transform (DPSIFT). The
proposed method takes less time for significant interest points detection, descriptors

extraction, and descriptor comparison for face recognition. The proposed method is tested with FJU, YouTube celebrity, and MIT-India face video datasets. It was found that the proposed DPSFIT method gives better results in terms of accuracy.

Acknowledgment The authors thank BRNS-BARC for support in the successful completion of the project work titled "Face Recognition System using Video Analytics."

References

1. Harris, C., Stephens, M.J.: A combined corner and edge detector. In: Alvey Vision Conference, 147–152 (1988)
2. Lowe, D.G.: Distinctive image features from scale-invariant keypoints. Int. J. Comput. Vis. (2004)
3. Bay, H., Ess, A., Tuytelaars, Tinne, Van Gool, Luc: Speeded up robust features. Comput. Vis. Image Underst. **110**, 346–359 (2008)
4. Viola, P., Jones, M.J.: Robust real-time face detection. Int. J. Comput. Vis. 57(2), (2004)
5. Cootes, T., Edwards, G.J., Taylor, C.J.: Active appearance models. IEEE Trans. Pattern Anal. Mach. Intell. 23(6), 681–685 (2001)
6. Beumer, G., Tao, Q., Bazen, A., Veldhuis, R.: A landmark paper in face recognition. In: 7th International Conference on Automatic Face and Gesture Recognition (FGR-2006). IEEE Computer Society Press (2006)
7. Cristinacce, D., Cootes, T.: Facial feature detection using adaboost with shape constraints. In: 14th Proceedings British Machine Vision Conference (BMVC-2003), 231–240 (2003)
8. Erukhimov, V., Lee, K.: A bottom-up framework for robust facial feature detection. In: 8th IEEE International Conference on Automatic Face and Gesture Recognition (FG2008), 1–6 (2008)
9. Cristinacce, D., Cootes, T., Scott, I.: A multistage approach to facial feature detection. In: 15th British Machine Vision Conference (BMVC-2004), 277–286 (2004)
10. Uricar, M., Franc, V., Hlavac, V.: Detector of facial landmarks learned by the structured output SVM, VISAPP '12: Proceedings of the 7th International Conference on Computer Vision Theory and Applications (2012)
11. Uricar, M.: Detector of facial landmarks, Master's Thesis, supervised by V. Franc (2011)

Registration of Range Images Using a Novel Technique of Centroid Alignment

Parama Bagchi, Debotosh Bhattacharjee and Mita Nasipuri

Abstract Here, the problem of 3D face registration across poses is addressed using the concept of alignment of principal components. First the alignment is done based on some coarse registration and then fine registration is computed using Iterative Closest Point. The registration scheme used is novel because for computing the coarse registration the distance between the centroids has been used. Subjective evaluation of registered images shows excellent registration.

Keywords Range image · ICP · Registration

1 Introduction

Recognition of human faces has been an important field of research. Past few years have found 3D face recognition to be more important than 2D face recognition, because of its enormous capability to handle pose, expressions, and occlusions [1–3]. Earlier, the problem of face recognition in humans had mainly relied on 2D images. The approach had certain limitations. Moreover, the information of a 2D face is often confined to pixel information only, thus making it difficult to assess the role of 3D shape processing. Also, a 3D face has more views that are necessary to

P. Bagchi (✉)
Department of Computer Science and Engineering, RCC Institute of Information Technology, Beliaghata, Kolkata, India
e-mail: paramabagchi@gmail.com

D. Bhattacharjee · M. Nasipuri
Department of Computer Science and Engineering, Jadavpur University, Kolkata, India
e-mail: debotosh@ieee.org

M. Nasipuri
e-mail: mitanasipuri@gmail.com

register for effective face recognition. The main characteristics of the present work have been enlisted below:

1. A new approach has been attempted which registers 3D faces based on their centroids.
2. This approach of centroid-based registration works well on poses.
3. The present method of registration is fully automatic and less complex, unlike landmark-based registration methods.

In Sect. 2, a literature survey has been given on 3D face registration. The present method has been discussed in Sect. 3. A comparative analysis with other methods existing in the literature has been demonstrated in Sect. 4. In Sect. 5, conclusion and future scope have been enlisted.

2 Literature Survey on 3D Face Registration

Illumination, expression, occlusions, and pose easily affect the performance of recognition rates in case of 2D face images because, 2D face images are incapable of capturing the depth information that is usually found only in case of 3D face images. Nowadays, it is possible to overcome these problems by 3D technology. Here, a survey is presented, which aims to address the various challenges and issues challenged by different 3D face registration techniques. Image Registration is defined as the process to align two different point clouds to each other. One point cloud is selected to be the reference, i.e., in frontal pose, and the other point cloud is the non-frontal, which is to be registered to the frontal point cloud. The process of registration continues till a minimum distance is attained between the frontal and the registered point cloud. There are two broad classifications of face registration:

(1) Coarse Registration Methods
(2) Fine Registration Methods

(a) **Coarse Registration Methods:** This form of registration could be categorized into the following:

  (i) Point Signature—This [4] descriptor is only for finding correspondences between two frontal and unregistered point clouds. The process is quite fast. However, the algorithm is very complex, and it is the main drawback of this algorithm.
  (ii) Spin Image—This [5] is a two-dimensional visualization. Initially, the method was used for recognition. One problem of the method is the resolution of the image. So, a filter is used, in this case, to remove false triangles and to give a smooth finish to the interpolated image.
  (iii) Principal Component Analysis—PCA [6] is usually used as a popular method for dimensionality reduction. This process is used for the alignment

of two point clouds. The disadvantage of principal component analysis is to cope with surfaces containing symmetries.

(iv) Algebraic Surface Model—Here [7], at first, from the two point clouds two implicit polynomial models are determined based on least squares method. However, this method is far better than (i) and (ii) above because it requires less time. The model is computed on the basis of a function of the distance between polynomial models and the points.

(b) **Fine Registration Methods:** This form of registration is used when a coarse registration has already been computed. The fine registration is computed based on an initial guess to converge to an optimal solution. The optimal solution is that the distance function between the two point clouds must be minimized after fine registration. Following are the categories of fine registration:

(i) Iterative Closest Point
(ii) Matching Signed Distance Fields
(iii) Procrustes Analysis.

Fine registration methods could be categorized into the following:

(a) ICP—Besl and McKay first presented the ICP [8] method. Here, the main aim is to obtain a solution in terms of fine registration between two different point clouds. In this case, an initial estimation between the two different point clouds is known, and so, all the points are thereby transformed to a common reference frame that is the frame for the registered image.

(b) Matching Signed Distance Fields—The method [7] is based on matching of signed distance fields and that is a multiview type of registration. First, all views are aligned with a common system based on an initial estimation of motion. Then, some key-points are generated on the three-dimensional mesh grid. Then, the closest key point would be searched to find out the necessary translational and rotational parameters required for the purpose of registration.

(c) Procrustes Analysis—Landmark forms the basis of the registration technique called Procrustes [8] analysis. For the alignment of two 3D shapes, landmarks are essential. So, the comparison of two shapes forms the basis of Procrustes analysis. Generalized orthogonal Procrustes analysis (GPA) is the registration of "k" sets of configurations.

3 Present Method

A depth map image (also called as a range image) is a 2.5D image. The speciality of the image is that at each (x, y) position of the image depth values of the image is stored, i.e., how far from the camera each point is situated. The 3D face registration is at this moment defined to be the process, by which, an unregistered point cloud is aligned to a registered or a frontal point cloud.

3.1   3D Face Registration

Face registration [1, 2] is defined to be the process to align an unregistered face image to a frontal face image. Registration requires some landmark points. So, registration is necessarily a transformation, which would align an unregistered point cloud to a registered point cloud (Fig. 1).

Face registration is important both in case of 2D and 3D face recognition. For registration, it is necessary to have a set of landmarks. For registration, translation of a probe image to a gallery image is necessary. The probe image is the unregistered point cloud. The gallery image is the image to which the probe image has to be registered. Point cloud is thus a 3D view of an image. The process of image registration can be of three different types [9]:

(i)   Registration of one face to another.
(ii) Registration to a fixed face model.
(iii) Registration to a coordinate system using facial landmarks.

The first approach registers two point clouds with an iterative procedure. One of the point clouds is the unregistered image while the other is the frontal image to which the unregistered image is to be registered. The second approach deals with the registration of an unregistered point cloud to a model that has been learned from a training set. Registration to a coordinate system using facial landmarks requires a mapping of the current landmarks of the facial system to the intrinsic coordinate system and thereby computing necessary translational and rotational parameters. The present approach may be visualized in the form of a block diagram in Fig. 2.

The following are the steps of the proposed algorithm:

(i)   **2.5D Range Image Acquisition:** The images acquired or downloaded are in the form of a Virtual Reality Modeling Language (VRML) file. The 3D data points to be visualized should be in the mesh format.
(ii) **3D Face Registration by ICP [5, 6]:** ICP is a method to register an unregistered face with a mean face template. In the present work, a simpler version of the ICP algorithm has been used. The ICP algorithm implemented in the present

(a)   **(b)**

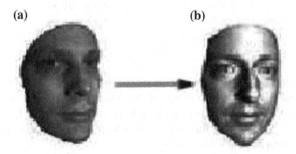

Fig. 1 Image for **a** Unregistered Image **b** Registered Image of a person

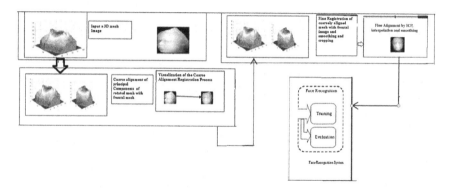

Fig. 2 Diagram showing the proposed approach (Range images are shown besides the 3D mesh for visualization purpose only)

work inputs two 3D mesh images, one image in frontal pose and an unregistered 3D mesh image from the 3D database, and tries to find a closest match between the two oriented models to find correspondences between the two. At first, the smallest average distance between the frontal and the unregistered model is used as an input for a coarse registration, after which a fine registration is sought.

ICP uses a transformation using a combination of:

a. PCA

b. SVD

At first, a coarse transformation is computed by aligning two point clouds based on the basic initial estimation of the closest points. Then the principal components are extracted. Next, the principal components are aligned using a fine transformation, for which it is necessary to compute the rotation matrix, which is required to register the unregistered point cloud to the frontal one.

- To perform the coarse registration, it is necessary to input the distance between corresponding points. The distance between the corresponding points (i.e., between the centroids) is passed as an input to the ICP algorithm, for the purpose of an initial coarse registration, after which the fine final registration is sought.
- When the distance between the closest points is passed to the ICP algorithm, the first initial transformation is found between the principal component of the unregistered and the mean frontal point cloud using reduction by principal component analysis.
- Next, a fine registration transformation is sought by aligning the two principal components and finally the 3D mesh is converted to the range image.
  The main contribution of the present work lies in the fact that, at first the smallest distance has been found between the two models for coarse registration. This distance is the distance between the centroids of the two point clouds.

Algorithm 1
ICP_Centroid_Align(model X, object P) //To Align Object P
to model X
Input: Initialize by finding the smallest correspondence
between each point on the three-dimensional surface between
principal components of the unregistered image and frontal
image. This smallest distance is taken to be the distance
between the centroids of the two point clouds.
Output: Registered Image
Coarse Registration:

1. At first, compute the distance between the centroids of the
   registered and the unregistered point clouds and save the
   distance in a variable that called as cen.
2. Roughly align the principal components of the unregistered mesh to that of the neutral mesh by applying steps 2.1
   to 2.3

   2.1 Compute the closest points in the unregistered mesh and
       find out the rotational and translational parameters
   2.2 Discard points in the unregistered meshes that are
       greater than cen.
   2.3 Find out the translational and rotational parameters by
       matching the closest points of the unregistered mesh
       and the neutral frontal mesh to which the unregistered
       model is to be registered
   2.4 Execute steps 2.1 to 2.3 for a fixed number of iterations

Fine Registration:
Input: Principal components of the coarsely registered
mesh X, the neutral frontal mesh P.
Output: The final fully registered mesh.

3. Repeat steps 4 to 5 for a specified threshold.
4. For all points that belong to P find closest x ε X.
5. Transform each point of the coarsely aligned point cloud X
   to minimize distances between P and X.
6. Output the registered image X

Figure 3 shows some images from GavabDB database, which has been registered using centroid alignment technique for poses. Figure 4 shows the corresponding principal components for the registered faces.

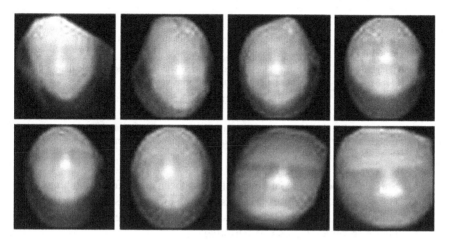

Fig. 3 Some registered face images from GavabDB database with pose variations

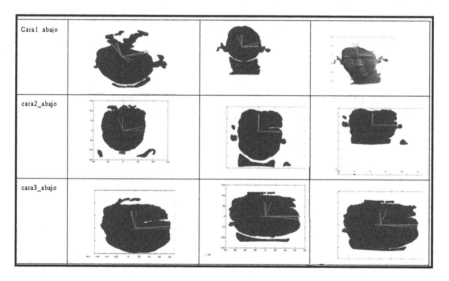

Fig. 4 A snapshot of the registration technique applied: *first column* showing principal components of unregistered mesh, *second column* showing the principal components to which the unregistered point clouds is to be registered and the *third column* shows principal components of the final registered image

(iv) **Noise Removal:** After registration of 3D mesh images have been done, the surfaces of 3D mesh images are processed for noise removal. For this, two steps are followed:

    (a) Interpolation: Here, entire 3D mesh image is preprocessed using an interpolation [4, 10] technique namely the trilinear interpolation.

Table 1 A comparative analysis of present technique over other methods existing in the literature

Sl. no	Name of the method	Database used	Novelty of technique used
1.	P. Bagchi [11]	Frav3D	Registration was performed by calculating the angle of pose variation.
2.	B.B. Amor [8]	Multiview 3D Face Database	Small poses registered using ICP
3.	Present method	GavabDB	Automatic registration using ICP using Euclidean distance between centroids

(b) Surface smoothing: Gaussian filter has been used for smoothing here. Gaussian smoothing uses a 2D distribution as a point-spread function, and convolution achieves this. Once a kernel has been calculated, then Gaussian smoothing can be performed.

(v) **3D Face Recognition:** The final stage, in the face recognition process, is feature extraction and classification. Here in this proposed work some efficient feature detection and extraction scheme is be used for final recognition preferably WLD feature extraction technique.

4   Comparative Analysis of the Present Approach with Other Methods Existing in the Literature

Table 1 depicts the manner in which the present method of face registration outperforms the already existing methods in the literature.

Analysis: In Table 1 some of the most recent works on ICP and some other methods are enlisted. Table 1 clearly shows that the present method of 3D face registration outperforms all the other methods present in the literature. Most methods that are present in the literature have registered small poses using ICP, but here the present method is quite an innovative one because it registers extreme poses too using ICP by centroid alignment.

5   Conclusion and Future Scope of the Work

The proposed work demonstrates the fact that registration is an essential transformation used for ill-posed subjects. In this work, ICP has been utilized in an entirely novel way, and the method is proposed to register extreme poses also. So, here, in

this proposed work, 3D face registration, its interpretability is expressed in registering extreme poses. Registration is a transformation that is used as a tool to
convert an ill-posed object to its frontal one. Once a subject is registered in a frontal
pose, after that, effective recognition could be done. Though the present method has
been tried on GavabDB dataset, the same is proposed in future to be tried on some
other existing datasets too. There are many advantages of 3D face recognition over
2D face recognition, and registration, as well as recognition, is an important field of
research work. As a part of future work, a proposal is being made to register and
recognize 3D faces across extreme poses by registering input face images using the
registration technique mentioned in this present work.

Acknowledgement   The work has been supported by the grant from DeiTy, MCIT, Govt. of India.

References

1. http://bosphorus.ee.boun.edu.tr/HowtoObtain.aspx
2. http://www.gavab.etsii.urjc.es/recursos_en.html#GavabDB
3. Sangineto, E.: Pose and expression independent facial landmark localization using dense SURF and the hausdorff distance. In: Pattern Analysis and Computing Vision (PAVIS), Genoa, Italy (2013)
4. Chua, C., Han, F., Ho, Y.K.: 3D human face recognition using point signature. In: International Conference on Automatic Face and Gesture Recognition (2000)
5. Li, Y., Smith, A.P, Hancock, E.R.: Face recognition with irregular region spin images. Lecture Notes in Computer Science, vol. 4522, pp. 730–739
6. Russ, T., Boehnen, C., Peters, T.: 3D face recognition using 3D alignment for PCA. In: IEEE Computer Society Conference on Computer Vision and Pattern Recognition (2006)
7. Tarel, J., Civi, H., Cooper, D.: Pose estimation of free form 3D objects without point matching using algebraic surface models. In: IEEE Workshop on Model Based 3D, (1998)
8. Amor, B.B., Ardabilian, M., Chen, L.: New Experiments on ICP Based 3D Face Recognition and Authentication. In: 18th International Conference on Pattern Recognition (ICPR) (2006)
9. Spreeuwers, L.: Fast and Accurate 3D Face Recognition. Int. J. Comput. Vis. **93**(3), 389–414 (2011)
10. Akima, H.: A method of bivariate interpolation and smooth surface fitting for values given at irregularly distributed points. In: ACM, TOMS (1978)
11. Bagchi, P., Bhattacharjee, D., Nasipuri, M., Basu, D.K.: A method for nose-tip based 3D face registration using maximum intensity algorithm. In: IEEE International Conference on Computation and Communication Advancement (2013)

Part II
Software Engineering

An Investigation into Effective Test Coverage

Debashis Mukherjee and Rajib Mall

Abstract Metric for coverage test addresses certain structures of the test program to infer on the test complexity. In our work we explore a popular metric in unit testing namely definition-use, and analyse the evolution of test criteria defined within *du* testing. We analyse examples to describe subsumption and ordering in the set of criteria, and strength and weakness of one criteria with another. We examine a test metric to infer on coverage and adequacy of testing achieved through it using Weyuker's set of properties. We enumerate a list of desirables those we thought contributed the effective structure in test coverage metric.

Keywords Test coverage · Control flow graph · Data and control dependence · Dataflow testing and properties

1 Introduction

The objective of a coverage metric is to quantitatively measure the thoroughness of software testing. The requirement for quantitative measurement of test coverage is to determine whether a given set of test cases achieve adequate amount of testing. This includes the unit test any incremental change with a least redundancy to changes introduced to the original test set, and regression testing [1, 2] of subset of the system in functional and structural dependence with the set of program statements in the change at the former; and towards an equivalence with best practices and established axioms later [2–4].

D. Mukherjee (✉) · R. Mall
Department of Computer Science and Engineering, Indian Institute of Technology Kharagpur, Kharagpur 721302, India
e-mail: debashis_mukherjee@yahoo.com

© Springer India 2016
R. Chaki et al. (eds.), *Advanced Computing and Systems for Security*,
Advances in Intelligent Systems and Computing 396,
DOI 10.1007/978-81-322-2653-6_7

1.1   Definition-Use (Du-Paths) *Path Coverage*

A large body of the literature exists on test coverage and metrics [3, 5–7]. Every test coverage is based on some program model. One of the earliest proposed program models is control flow graph (CFG) of the program execution from start to stop of a test run, with the program statements at the nodes of the graph. Along the control path, the data flow from one node to another is tested in data flow testing [4, 8, 9]. Testing is measured between a pair of nodes in each time. The node where the data is defined is known as *def* node, and the node where the data is used is known as *use* node. The testing of data flow between pairs of *def* and *use* nodes is mainstream in *definition-use (du)* testing.

Paths are defined from each definition node to one or all uses of definitions in sets or collections of paths in *du-coverage*, where each sub-path from a *def* to *use* node is of the type of simple cycle path, and a sub-path with no intervening re-definition of the variable after the *def* node to the *use* node (*def-clear*). The *use* node for computation is known as *c-use*, and when used in branch statements as predicate, is known as *p-use*. *All-defs-paths* is the sets of paths containing cycle free *def-clear* sub-paths from each *def* node to at least one *use* node from the definition in all *def* nodes in the *du-pairs*. *All-uses-paths* is the sets of paths containing cycle free *def-clear* sub-paths from each *def* node to all *use* nodes reachable from the definition in all *def* nodes of the *du-pairs*. *All-du-paths* is the sets of paths containing cycle free and simple cycle *def-clear* sub-paths from each *def* node to all *use* nodes reachable from the definition in all *def* nodes of the *du-pairs*. Set of the all possible paths on any pair of nodes on the control flow graph is called as *all-paths*, and amounts countably infinite in number in presence of cycles and loops in the control flow graph.

The sets of *du-paths* in *definition-use* metric, forms a subsumption structure, represented by partially ordered relation. Statement coverage refers to execution of each of the program statements by a set of test cases, or test suite. Decision or branch coverage refers to execution of each of the boolean outcome (*true* and *false*) of each of the conditions in the program statement. CFG-edge coverage refers to execution of each of the ordered pair of nodes or the edges of the control flow graph model of the program. Statement coverage is subsumed in branch coverage, branch coverage is subsumed in CFG-edge coverage, branch or CFG-edge coverage is subsumed in *all-uses-paths* coverage, *all-defs-paths* coverage is subsumed in *all-uses-paths* coverage, *all-uses-paths* coverage is subsumed in *all-du-paths* coverage, and each coverage metric in subsumed in *all-paths* coverage. The subsumption hierarchy defined by *du*-coverage [4] is shown in Fig. 1.

The program model for *du*-coverage is the control flow graph and a notion of *du-pair* describing temporal data dependence between two nodes at an instant on a test run on the graph. A data dependence is exercised on CFG, is utilised to describe sequence of definition-use pairs, and definition by definition pairs to represent dynamic interactions of test scenario and test context, and are named as *k-dr* interaction [8] and DC and ODC of test context [9] respectively.

Fig. 1 Subsumption of test
coverages [4]

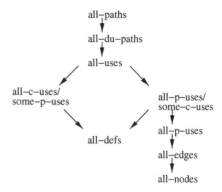

1.2   Enhancement in Coverage on Path-Based Metric

It is possible to define more paths as sets of paths by building on the coverage
inherent in *du-path* coverage. Data flow between a pair of nodes with one *def* node
on a variable to one of its *use* nodes on a path, or sets of paths from one *def* node to
all of its *use* nodes along all possible paths is used as criteria in *du-path*. The
coverage criteria can be enhanced with additional paths defined as below: A *mul-
tiple-definition-to-use-path*, is a path to a *use* node, from *def* node of each variable
to the *use* node on a simple cycle path in test from *start* to *end* node in the CFG.
A *definition-to-definition-use-path*, is a path through a *def* node on a variable to an
intermediate *use* node on the variable, followed by definition of another variable or
re-definition (*undef*) of the variable by a *def* at the same intermediate node, further
followed by, use of the intermediate variable at a *use* node observed in the
sequence. The above criteria, may be applied in sets of paths in conjunction with
all-paths in the type, applies more constraint to paths, for enforcing test paths in
specification to pass through required *def* and *use* nodes in the program.

1.3   Dependence Paths Coverages

A graph defined on the nodes of the control flow graph, and edges defined by
relation due to control dependence and data dependence on a single procedure
program is called program dependence graph (PDG). An extension of PDG to
model the semantics of procedure calls, containing multiple dependence graphs for
procedures other than the main method with additional edges to interconnect the
graphs to the main program dependence graph is called system dependence graph
(SDG).

Coverage metric defined on PDG is known as dependence path coverage.
A sub-graph of PDG, containing only the data dependence edges is called data
dependence graph (DDG), and those with only the control dependence edges is

called control dependence graph (CDG), used together with PDG is defining data and control dependence path coverages, respectively. Dependence path coverage is due to augmentation of coverage criteria of shorter/disjoint data dependence paths, to longer and non-increasing number of paths in the coverage criteria, with addition of sequences of control dependence paths/edges [10]. It could be shown that *du-paths* coverage is subsumed by data dependence path coverage, and data dependence path coverage is subsumed by dependence path coverage. Dependence path coverage applies non-intervening data definition sub-paths on paths of CFG. Hence, dependence path coverage cannot subsume *all-paths* coverages.

We have so far discussed preliminaries of test coverage and two specific metric—*du-path* and *dependency path* coverage. The rest of paper is organised as follows. We describe the very basic concepts and investigate some topics in subjective interest in the context of test coverage in Sect. 2. In Sect. 3, we enumerate a list of desirables from a metric.

2 Basic Concepts

2.1 Control Flow Graph

The control flow graph (CFG), $G_C(V, E)$ for a program β is a directed graph in which the vertices $v \in V$ represent the statements in a program and the edges $e \in E$ represent the control flow between two statements in succession. A path in a control flow graph is a sequence of edges. A CFG consists of two additional nodes, *start* and *stop* node, representing entry to the system, and exit from the system respectively. All complete test path starts from *start* node and ends at the *stop* node of the CFG. Figure 2a shows an example procedure and its corresponding control flow graph.

Domination A vertex, v_i is said to *dominate* (or predominate) a vertex v_j in a control flow graph, if every path from the *start* to vertex v_j passes through v_i. A vertex v_m is said to *post-dominate* a vertex v_n, if every path from v_m to the *stop* vertex passes through v_n. In the CFG in Fig. 2, vertex 9 is dominated by vertex 8; vertices 6 and 7 is dominated by vertex 5; vertices 2, 3, 4 and 5 dominate vertex 8; vertices 6, and 7 post-dominate vertex 5; and vertex 9 do not post-dominate vertex 8.

Control dependence A vertex v_j is *control dependent* on vertex v_i on the control flow graph (CFG) G_C iff, there exist a path p from v_i to v_j; vertex v_i dominates and does not post-dominate vertex v_j. In the CFG in Fig. 2, vertex 9 is control dependent on vertex 8; vertices 6 and 7 is control dependent on vertex 5. Please note that vertex 9 does not post-dominate vertex 8; whereas, vertices 6 and 7 does to vertex 5.

Data dependence A vertex v_j is *data dependent* on vertex v_i on the CFG G_C of a program β, and x be a variable in β iff, vertex v_i defines x; vertex v_j uses x; and there exists a path from v_i to v_j along which there is no intervening definition of x. In the

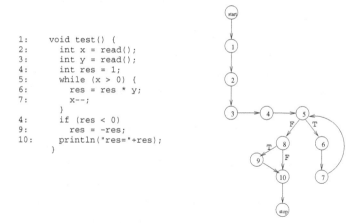

```
1:      void test() {
2:          int x = read();
3:          int y = read();
4:          int res = 1;
5:          while (x > 0) {
6:              res = res * y;
7:              x--;
            }
4:          if (res < 0)
9:              res = -res;
10:         println("res="+res);
        }
```

Fig. 2 An example procedure and its control flow graph representation

example in Fig. 2, vertex 6 is data dependent on vertices 3, 4 and 6 (itself); and vertex 10 is data dependent on vertices 4, 6 and 9.

2.2   Definition-Use *Graph*

A *definition-use* graph (DUG) is an annotated control flow graph in which the nodes are annotated with two sets of variables: a set of variables that are defined; and a set of variables that are used at the statements represented by the nodes. DUG is also referred to as *def-use* graph or *def/use* graph, and is formally defined on control flow graph G_C for program β and set of vertices V, in the following.

def(v_i), use(v_i) For each vertex $v_i \in V$ in G_C, $def(v_i)$ denotes the set of variables defined at the statement represented by v_i, and $use(v_i)$ denotes the set of vertices used at that statement.

def(p)   Let p = $(v_i, v_{i+1}, ..., v_j)$ be   a   path   in G_C,   then $def(p) = def(v_i, v_{i+1}, ..., v_j) = \bigcup_{v_k \in p} def(v_k)$.

Def-Use graph A *def-use* graph for a program β is defined as a quadruple $G_D = (G_C, S, def, use)$, on control flow graph G_C defined for program β; S is finite number of symbols called variables that are used in the program; and $def:V \rightarrow 2^S$, and $use:V \rightarrow 2^S$ as functions. A *definition-use* graph of example program in Fig. 2 is shown in Fig. 3.

def-clear Path Let x be a variable in a program β, and $p = (v_i, v_{i+1}, ..., v_l, v_j)$, $k > i$, $k \leq l < j$ be a path in G_C, then the path p is said to be a *def-clear* path from v_i to v_j with respect to x if $x \notin def(v_k, ..., v_l)$.

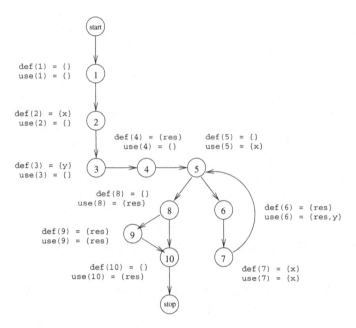

Fig. 3 Definition-use graph for the example program in Fig. 2

2.3 Concept Learning

A perspective on machine learning involves searching a large space of hypotheses and infers boolean-valued function from examples of its input and output in an inductive learning to approximate a target function [11]. A boolean hypothesis h_j is said to be *more_general_than_or_equal_to* boolean hypothesis h_k ($h_j \geq_g h_k$) iff, ($\forall x \in X$) $[(h_k(x) = 1) \rightarrow (h_j(x) = 1)]$, where instance x in X, and hypothesis h in H. Maximally specific hypothesis is searched (aka *Find-S*) using *more_general_than* partial ordering consistent with the examples observed, and is said that a hypothesis "covers" a positive example if it correctly classifies the example. The version space $VS_{H,D}$, with respect to hypothesis space H and *consistent* (in absence of satisfactory boolean function at its disposal) set of examples D, is represented as $VS_{H,D} \equiv \{h \in H | Consistent(h, D)\}$.

The concept of a program instance in Fig. 5 is realised as shown in Fig. 4a, represented of instances of test paths for test input $\langle 7,1 \rangle$, $\langle 6,1 \rangle$ and $\langle 8,1 \rangle$ (with equivalent representative set of nodes of corresponding CFG paths of 4a as {1,2,3,4,5,6,7,8,9}, {1,2,3,4,5,6,4″,5″,6″,7,8,9} and {1,2,3,4′,7′,8′,9′}, respectively) on consistent hypothesis under program model of program dependence graph (PDG) (dependence path coverage [10]) in Fig. 4b to some well-formed version in Fig. 4c.

(a)   **(b)**   **(c)**

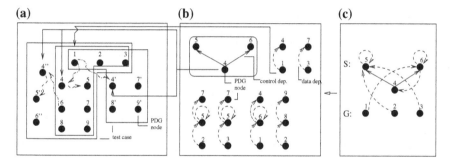

Fig. 4 A test concept learning scenario: **a** test set instances, **b** hypothesis space, **c** concept/version representation of a while loop of sample program in Fig. 5a

Fig. 5 An example instance to learning. **a** Sample program. **b** Control flow graph (CFG)

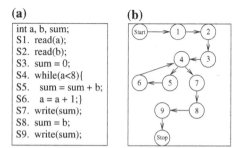

(a)

```
int a, b, sum;
S1.  read(a);
S2.  read(b);
S3.  sum = 0;
S4.  while(a<8){
S5.     sum = sum + b;
S6.     a = a + 1;}
S7.  write(sum);
S8.  sum = b;
S9.  write(sum);
```

(b)

2.4 Abstract Interpretation

An unique test run through a program may contain cyclic sub-paths on the CFG representation. Such paths may not be *du-path*, or may not even possible to be included (enforced/covered) into the set of paths belonging to test cases (as test paths defined on the CFG) and included into the test coverage using coverage criteria of dependence coverage, under static constraints of dependency paths coverage in program model of program dependence graph (PDG).

Static analysis of the program is required in such cases using data flow equations on the control flow graph (CFG). Coverage of each static blocks (nodes or statements on CFG) and branches, needed to be correlated to include these to test scenarios covering for statements reachable, into statically analysed dependency paths coverage.

The paths through a program might be of the type feasible or infeasible, and is determined by the branch predicates at runtime. Certain subsets of such paths can be interpreted through static analysis of data flow equation and reachability analysis of regions or labels or locations in the CFG. The range of values taken by the variable or propagation of interval values determines the choice of branch in the dynamic case. A test metric is required for attempts to support in the choice and inclusion of test paths precisely. For example, in Fig. 6, it is possible to reach to

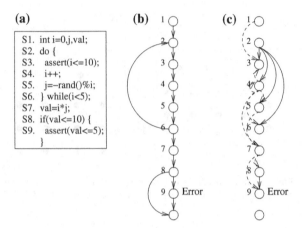

Fig. 6 Path with cyclic sub-paths in critical test. **a** Sample program. **b** CFG. **c** PDG

```
S1. int i=0,j,val;
S2. do {
S3.   assert(i<=10);
S4.   i++;
S5.   j=−rand()%i;
S6. } while(i<5);
S7. val=i*j;
S8. if(val<=10) {
S9.   assert(val<=5);
    }
```

sub-path 8-9 and 6-2-3 as incorrect and potentially failing to match a specification of correctness through static analysis, however, creation of a test run in support to a proof might not be trivial. The sub-path 8-9 could appear in a feasible path at random, in the path 1-[2-3-4-5-6 5 times]-7-8-9.

2.5 Reliability of Networks

The edges of control flow graph (CFG) or the dependence graphs (PDG, SDG), could be labelled with responses from a system test coverage on test suites, assuming that the system resembles a network whose reliability would be reflected as weights against tolerate the faults. The nature of statistical data is whether a passing or failing run of the tests include or exclude a node or vertex of the graph [12], consequently those edges explored to reach the vertices.

2.6 Acceptance Tests

A system is tested for a range of validations of output on given input, assumed as acceptable or as specified in its acceptance testing. Effectiveness of a test depends on narrowness and width of the range, alternatively termed as sensitivity and specificity [1, 13] of the test, respectively. Program model of the statements or the system may be useful in acceptance test coverage structure on isolation specificity in granularity ranging modules in black-box testing to statements in the program in white-box testing [1].

2.7 Program Slicing

The value of a variable at program point is influenced by only a subset of all possible statements executed before the statement. Similarly, a variable at a program point influences only a subset of all statements. These two types of analyses are referred as forward and backward slice, respectively in the literature on program slicing [14]. The slice of program computed statically on the entire program, is known as static slice and when computed with respect to statements executed in a test run (or path) is known as dynamic slice. The criteria for slice are comparable with control dependence and data dependence, and transitive closure of the dependences at a program point forms the corresponding slice [10]. Slicing when applied to extract relevant statements that have been transformed from the original set of statements is known as amorphous slice. Slicing is used to reduce complexity of program analysis at a program point by lowering redundancy and increasing parallelism.

2.8 Complexity Metric

Notion of cyclomatic complexity is an account of graphical model of flow of control with respect to constructs in programming languages in structured programs. Complexity for a simple program is estimated as number of conditional statements, alternatively as the number of complete paths through the program to test the program at a minimum [6]. The number of *du-path*, effectively indicate the required number of structural paths to test the program adequately with data flow as constraints into consideration, with one data constraint exercised per test run on per path. In *dependency path*s, an adequacy is determined in relation to the exercising one or more data and control dependence per test run on per test path. Complexity as number of test runs to explore to test a program could reduce summarily lower in dependency path coverages than *du-paths* coverage; and cyclomatic complexity would be almost equal and least in most cases with multiple unstructured branching.

3 Desirables of the Test Metric

A test metric requires to exercise all the corners of bounded region in the program those presumed as adequate in its model to test its subject, or set of program statements, and could be in response to queries as follows:

3.1   Set of Test Paths Included

A metric should be able to include as many paths in the program defined on the model. On a control flow graph, a set of all such paths uniquely and may not as simply, could be described as *def-clear* and simple cycle complete paths through the program statements. A metric describe all paths through program statements defined by the nodes reachable through connectivity in a graph model.

3.2   Definition of Underlying Hypothesis and Structure

Every program model defines a set of constraints or hypothesis, to represent the coverage that it targets to describe. An implicit structure helps in describing the constraints to model a coverage problem. A general to specific structure [11] is natural in the representation of the space of hypothesis.

3.3   Extendability, Subsumption, Relation on Established Metric

Coverage metrics extend and build on established metric to an available extent, and tries to intervene at best possible points in the subsumption structure. The definition of criteria in the coverage metrics aim to attain the partial order and overcome any weakness of criterion in immediate lower through the criterion in its upper using an order.

3.4   Hypothesis Induced Per Path in the Tests Run, and/or Paths Induced Per Hypothesis

Specificity of an included path, to specificity of a constraint finds necessary in isolation of critical components for attention on any cause and its effect. The effectiveness of a coverage metric may be attributed by fractional effect on these ratios when contributed in its model. For example, the *du* criteria *uses*, extends description through set of specific *def-clear*s, and the *all-uses* addresses almost every tests which are not simple cycle. Specificity of tests required to simple cycle path is increased through *all-du-path*s.

3.5 Compatibility with Specific Applications

Compatibility features in programming with high level language, in presence of arrays referencing, inter procedure calls and handling of exception conditions; and specific application goals in domains of fault localization, test for redundancies, parallelism, etc. are necessary deliverables of a test coverage metric to the front-end of validation/verification tools.

4 Weyuker's Properties and Dependence Path Coverage Metric

Table 1, describes properties of test adequacy [15] as realised in a dependence path coverage metric. Referring to program dependence graph (PDG) as the program model for dependence paths coverage, let us denote PDGs/dependence graphs

Table 1 Weyuker's properties and dependency metric

Weyuker's set of properties	Dependency metric
$(\exists P)(\exists Q)(|P| \neq |Q|)$, interpreted as "Class identifier" measure	Dependence path coverage admits the property, as two unequal PDGs, correspond to two distinct complexity measures, on trivial case
Let c be a non-negative number. Then there are only finitely many programs with complexity c, interpreted as "Size Independent" measure	There are only finitely many PDGs possible to be constructed, on sub-case where complexity correspond to functions mapping to integer from discrete graph structure, hence dependence path coverage admits the property
There are distinct programs P and Q such that $(|P| = |Q|)$, interpreted as "Indexing functions" measure	Two distinct programs, irrespective of their PDGs distinct, could correspond to complexity measures (as monotonic function from connectivity, vertices, edges, etc.on the graph, to numbers as non-negative integers) which may be equal, hence the dependence path coverage admits the property
$(\exists P)(\exists Q)(P \equiv Q \text{ and } |P| \neq |Q|)$, interpreted as "Interface sensitive and implementation insensitive" measure	Two unequal PDGs (consequently, corresponding to unequal complexities), may correspond to be equivalent programs considered on input/output, etc., hence dependence path coverage admits the property
$(\forall P)(\forall Q)(|P| \leq |P; Q| \text{ and } |Q| \leq |P; Q|)$, interpreted as "Interaction sensitive metrics"	Concatenation of two programs would correspond to a PDG, which would contain the PDG corresponding to the concatenating programs, as a sub-graph, and the property

(continued)

Table 1 (continued)

Weyuker's set of properties	Dependency metric
	holds, with the equivalent measure of complexity metric being "monotonic" on variables (as, vertices and edges on PDG)
$(\exists P)(\exists Q)(\exists R)(\|P\| = \|Q\|$ and $\|P; R\| \neq \|Q; R\|)$. and $(\exists P)(\exists Q)(\exists R)(\|P\| = \|Q\|$ and $\|R; P\| \neq \| R; Q\|)$., interpreted as "Interaction insensitive metrics"	Two different programs (such as, with renamed variables, or permuted statements, etc.) could correspond to PDGs which are equal. Concatenation of such programs, with a known program, could correspond to PDGs, which may not be equal, hence dependence path coverage admits the property
There exists programs P and Q such that Q is formed by permuting the order of the statements of P and $(\|P\| \neq \|Q\|)$, interpreted as "Building block count metrics"	If PDG $P_Q \langle V_Q, E_Q \rangle$ is obtained after permuting the set of program statements corresponding to PDG $P_P \langle V_P, E_P \rangle$, then $E_P \neq E_Q$, would imply $P_P \neq P_Q$, hence dependence path coverage admits the property
If P is renaming of Q then $(\|P\| = \|Q\|)$, interpreted as "Name sensitive metrics"	Assuming a renaming would only impact the program variables consistently in the language, with no impact to the structure of the program dependence graph, such that the PDG $P \langle V, E \rangle$ remains equal, hence dependence path coverage admits the property
$(\exists P)(\exists Q)(\|P\| + \|Q\| < \|P; Q\|)$.	Dependence path coverage admits the property due to the program model of PDG, as $\|V_{PQ}\| \geq \|V_P\| + \|V_Q\| + 1$, and $\|E_{PQ}\| \geq \|E_P\| + \|E_Q\| + 2$, as P_{PQ} would contain at least one more vertex and two more edges (corresponding to entry node representing control dependence, and two edges to at least to one statement in each set in the combined PDG $P_{PQ} \langle V_{PQ}, E_{PQ} \rangle$) over PDGs P_P, P_Q as sub-graphs in PDG P_{PQ}

corresponding to sub-program P as $P_P \langle V_P, E_P \rangle$, Q as $P_Q \langle V_Q, E_Q \rangle$, and a combined/concatenated program P;Q as $P_{PQ} \langle V_{PQ}, E_{PQ} \rangle$, in the rest of the tabulation, where notation V correspond to set of vertices and E correspond to set of directed edges (data and control dependence edges) for PDG $P \langle V, E \rangle$, in an usual sense.

5 Conclusion

Each of the metrics proposed in test coverage in the prior research has some shortcomings. In our work we investigate for a test metric that is based on some established program model, and is capable of addressing cases of programming language features spanning, global variables, arrays referencing and exceptions handling in structured programming languages and intended on object oriented software systems in the future.

References

1. Mall, M.: Fundamentals of Software Engineering. PHI learning pvt. ltd., New York (2009)
2. Zhu, H., Hall, P.A.V., May, J.H.R.: Software unit test coverage and adequacy. ACM Comput. Surv. (CSUR) **29**(4), 366–427 (1997)
3. Weyuker, E.J.: Evaluating software complexity measures. IEEE Trans. Softw. Eng. **14**(9), 1357–1365 (1988)
4. Rapps, S., Weyuker, E.J.: Data flow analysis techniques for test data selection. In: Proceedings of the Sixth International Conference of Software Engineering, Tokyo, Japan, pp. 272–277 (1982)
5. Chidamber, S.R., Kernerer, C.F.: A metrics suite for object oriented design. IEEE Trans. Softw. Eng. **20**(6), 476–493 (1994)
6. Nejmeh, B.A.: NPath: a measure of execution path complexity and its applications. Commun. ACM 188–200 (1988)
7. McCabe, T.J.: A complexity measure. IEEE Trans. Softw. Eng. 308–319 (1976)
8. Natofos, S.C.: On testing with required elements. In: Proceedings on COMPSAC'81, IEEE Computer Society, pp. 132–139 (1981)
9. Laski, J.W., Korel, B.: A data flow oriented program testing strategy. IEEE Trans. Softw. Eng. **9**(3), 347–354 (1983)
10. Mukherjee, D.: Dependency cov. criterion is testing. Technical report, submitted to CSE Dept., IIT Kharagpur (2015)
11. Mitchell, T.: Machine learning. McGraw-Hill, New York (1997)
12. Jones, J.A., Harrold, M.J., Stasko, J.: Visualization of test information to assist fault localization. In: Proceedings of the 24th International Conference on Software Engineering, ICSE '02, pp. 467–477. ACM, New York (2002)
13. Koren, I., Krishna, C.M.: Fault Tolerant Systems. Elsevier, San Francisco (2007)
14. Weiser, M.: Program slicing. In: Proceedings of the 5th International Conference on Software Engineering, ICSE'81, pp. 439–449. IEEE Press, Piscataway (1981)
15. Weyuker, E.J.: The evaluation of program-based software test data adequacy criteria. Commun. ACM **31**(6) 1988
16. Halstead, M.H.: Elements of software science. Elsevier, North-Holland (1977)
17. Myers, G.: An extension to the cyclomatic measure of program complexity. ACM SIGPLAN **12**, 61–64 (1977)
18. Chen, E.T.: Program complexity and programmer productivity. IEEE. Trans. Softw. Eng. 187–194 (1978)
19. Baker, A., Zwehen, S.: A comparison of measures of control flow complexity. IEEE Trans. Softw. Eng. 506–512 (1980)
20. Evangelist, W.: Software complexity metric sensitivity to program structuring rules. J. Syst. Softw. 231–243 (1983)

21. Kernighan, B., Plauger, P.: The elements of programming style, 2nd edn. McGraw-Hill, New York (1978)
22. Aho, A.V., Ullman, J.D.: The theory of parsing, translation, and compiling, vol. 1. Prentice-Hall, Englewood Cliffs (1972)
23. Beth McColl, R., McKim, J.C. Jr.: Evaluating and extending NPath as a software complexity measure. J. Syst. Softw. **17**, 275–279 (1992)
24. Lakshmanan, K.B., Jayaprakash, S., Sinha, P.K.: Properties of control-flow complexity measures. IEEE. Trans. Softw. Eng. **17**(12) 1991
25. Clarke, L.A., Podgurski, A., Richardson, D.J., Zeil, S.J.: A formal evaluation of data flow path selection criteria. IEEE Trans. Softw. Eng. **15**(11), 1318–1332 (1989)
26. Mund, G.B., Mall, R., Sarkar, S.: Computation of intraprocedural dynamic program slices. Inf. Softw. Technol. **45**, 499–512 (2003)
27. Mund, G.B., Mall, R.: Chapter 8: program slicing: the compiler design handbook: optimizations and machine code generation. CRC Press, Boca Raton (2003)
28. Schwarze Braunschweig, J.: An algorithm for hierarchical reduction and decomposition of a directed graph. Computing Springer, **25**, 47–47 (1980)
29. McConnell, R.M., de Montgolfier, F.: Linear-time modular decomposition of directed graphs. J. Discrete Appl. Math. Struct. decompos. width parameters graph label. **145**(2) (2005)
30. Najumudheen, E.S.F., Mall, R., Samanta, D.: A dependence representation for coverage testing of object oriented programs. J. Object Technol. **9**(4), 1–23 (2010)
31. Cousot, P., Cousot, R.: Abstract interpretation: a unified lattice model for static analysis of programs by construction or approximation of fixpoints. In: Symposium of ACM special interest group on automata and computability theory special interest group on programming languages (1997)

Resource Management in Native Languages Using Dynamic Binary Instrumentation (PIN)

Nachiketa Chatterjee, Saurabh Singh Thakur and Partha Pratim Das

Abstract Managed programming languages like Java and C# perform resource management as a part of their language specification. They use a runtime system like JVM or CLR for the management. In contrast native languages like C and C++, designed to provide strong foundation for programs requiring speed or tight coupling with operating system or hardware, are used with manual resource management. These do not require the runtime system. Naturally, it will be nice to have a managed layer for native languages which can be plugged in as and when we want to manage resources in any point of time during execution. In this paper, we present a GC Pintool which automates the garbage collection for C programs at run time using PIN (a framework for dynamic binary instrumentation). Efficacy of the GC Pintool has been tested over various benchmark C programs and our GC approach using PIN is found to be correct and precise.

Keywords Garbage collection · Memory leak · Dynamic instrumentation

N. Chatterjee (✉)
A. K. Choudhury School of Information Technology, University of Calcutta, Kolkata, India
e-mail: nachiketa.chatterjee@gmail.com

S.S. Thakur
School of Information Technology, Indian Institute Technology, Kharagpur, India
e-mail: saurabhjan07@gmail.com

P.P. Das
Department of Computer Science and Engineering, Indian Institute Technology,
Kharagpur, India
e-mail: partha.p.das@gmail.com

R. Chaki et al. (eds.), *Advanced Computing and Systems for Security*,
Advances in Intelligent Systems and Computing 396,
DOI 10.1007/978-81-322-2653-6_8

1   Introduction

The application programming languages available in market can be categorized into two types depending upon their style of execution and resource management. Most of the popular languages like Java, C#, etc., can automatically manage their resources such as memory, graphics wizard, etc. So that they usually are termed as *managed languages*. But, the managed languages need a runtime system for execution that adds additional overhead. In contrast, native languages like C and C++ facilitate users to write high performance and responsive applications with direct interaction with hardware resources. But, the user experiences the overhead to manage resources manually. With the speed and flexibility of C and C++ comes increased complexity along with the complications in memory management. Objects must be created and destroyed explicitly in program, and small mistakes in this process can cause severe complications.

Garbage collection (GC) is the most popular automated technique for memory management. A garbage collector detects objects that are not being used by the program and attempts to reclaim the memory (garbage) occupied by those objects. It liberates the programmer from the responsibility of taking care of dynamically allocated memory and is based on the following principles:

1. To identify the objects in memory that cannot be accessed any further, and
2. To destroy these objects and reclaim the memory used by them.

While garbage collection is used, certain categories of bugs, as described below, get eliminated or are substantially reduced.

1.1   Memory Leak

When an object becomes unreachable, the program fails to free its memory and leak occurs. This memory then becomes unavailable to the system. Series of memory leaks may cause a program to crash due to memory exhaustion. Even if not, then also it can have an adverse effect on performance. Chunks of allocated but unused memory cause fragmentation. It destroys the spatial locality and this can result in poor cache performance or an increase in paging.

1.2   Dangling Pointer

When the program holds more than one pointer to a memory, the memory is made free through one of them, and is then accessed by another, an illegal dereferencing happens for the dangling pointer. By the time of access, the memory may have been reassigned for some other use. Such dangling pointer access, therefore, may lead to unpredictable results.

Managed programming languages currently in use perform Garbage Collection automatically. So, it will be nice to have a managed layer for native languages which can be plugged in when we want to manage resources in any point of time during execution.

We organize the paper as follows. Section 2 describes about the various resource management techniques in brief. Section 3 discusses about the strategy of resource management using dynamic instrumentation. Section 4 explains about the details of implementation of our resource management framework; discusses its various modules and their implementation. The results for the test cases are presented in Section 5. In this section, we will discuss our test plan and the various benchmark codes that we have used to test the GC Pintool. Section 6 concludes and suggests some possible future research directions resulting from this work.

2 Resource Management Techniques

There are a couple of classical resource management mechanisms available in managed languages those that have been summarized as below. Also, we found some attempts of recourse management for native languages.

2.1 Classical Resource Management

In general, a garbage collection process involves three basic steps as shown in Fig. 1, i.e., scanning the root, marking, sweeping, and an optional step of compacting. There are two generic tasks that a garbage collector needs to perform

1. Distinguish between garbage and live objects and
2. Remove garbage to reclaim memory.

The garbage collection process is initiated with the root scanning that examines if a memory object includes any pointer. A reachability graph is then formed with

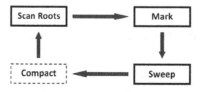

Fig. 1 Typical garbage collection cycle

the objects and the pointers to the objects. Naturally, one can traverse this graph to compute if an object is reachable. When an object is reachable (from the root) in this graph, the program can (potentially) access it by navigating the pointers to it and other objects. Hence, all reachable objects are detected and marked. Objects that are not marked are not reachable, and are garbage by definition [1]. One can then sweep (or walk) through the heap and deallocate the unmarked objects in the process of reclamation. Alternately, in some systems, compaction is used on the heap to narrow the gaps created by unmarked objects that are removed. Compaction minimizes external fragmentation [2] at the cost of object relocation. Recently, several garbage collection techniques have been proposed with their respective properties. They can be grouped as follows:

Reference Counting is one of the oldest resource management techniques, where every object has a counter holding the number of pointers that point to the object. The counter is incremented when a new pointer starts pointing to the object. When a pointer referencing the object is reassigned, the counter is decremented. The object becomes unreachable by the *mutator* [3] (executing program) as soon as the counter becomes zero. It is then returned to the free pool. The strength of this technique apart from its simplicity is immediate recovery of unreferenced memory objects [4]. However, this method cannot handle cyclic references as the two (or more) objects on the cycle are having nonzero counter value [5]. They are left unreclaimed.

Mark and Sweep GC, as the name suggests, first *marks* the reachable objects and then *sweeps* them. It adheres strongly to the two-phase abstraction of garbage collection. Marking is done recursively starting from the root set. For sweep, the heap is scanned linearly and all unmarked objects are reclaimed [3, 6].

Copying Collector copies reachable objects from one part of memory to another [3] and reclaims the garbage in the process.

Generational Garbage Collectors are based on *generational hypotheses* and attempt to improve performance by using the age of objects. The *weak generational hypothesis* professes that most objects must die when they are young [7]. In contrast, the *strong generational hypothesis* assumes that as an object becomes older, it becomes less likely to die. Generational collectors have been shown to generally outperform their nongenerational counterparts [8], and are today the most commonly used type of collectors for the majority of systems.

Incremental Garbage Collectors are meant to minimize the disruptiveness of collectors, specifically, those that have long pauses. For this, the collector is run in tandem with the application program so that it can gradually do its work [9].

2.2 Resource Management for Native Language

Two approaches for adding GC to C, namely *Conservative GC* [10] and *Precise GC* [11], have been proposed earlier.

Conservative GC attempted to automate GC for C program where the programmer needs to link with *Boehm GC Library* and use its allocator/deallocator functions. This GC technique operates roughly in four phases of the Mark-Sweep algorithm. But it falls short in identifying pointer variables in some cases where it is unable to determine whether a word is a pointer or wrongly assumes the dead pointers as roots and leaves objects indefinitely in the heap. For long-running programs like an IDE, a web server or an operating system kernel, *Conservative GC* does not perform well. While managing threads and continuations, it can potentially cause unbounded memory use due to linked lists [12]. The problem is caused by liveness vagueness [13], rather than type imprecision.

Precise GC [11] is an improvement over the conservatism of *Conservative GC* [10] in terms of memory and time. It uses Magpie and performs source-to-source transformation that rely on an ontology of objects including "root references in heap," "location of reference in every kind of object," and "types of objects in the heap." This results in overhead for the mutator besides requiring additional programmer effort and related complexities. For example, while using *Precise GC* for a C program, a pointer must never be extracted from a variable typed as long—not at least after a collection has taken place since the variable was assigned. However, compared to *Conservative GC*, *Precise GC* makes weaker assumptions on the compiler and the architecture. The original program is transformed to explicitly cooperate with the GC.

In this paper, we have investigated the solution to the problems of memory management in C programs. To overcome the shortcoming of *Conservative GC* [10] and *Precise GC* [11], we use dynamic instrumentation that does not need any modification in the source code as were required for both the above mentioned GC techniques. In this work, a GC Pintool has been developed which automates the process of garbage collection for C Programs. The tool addresses a wide range of memory management issues for C programs and efficiently improves the performance of C programs in context with the memory management.

3   Resource Management Using Dynamic Instrumentation

To design a pluggable managed layer for native languages, we have used a dynamic instrumentation framework called Pin [14]. It is a platform for runtime binary instrumentation of applications running on Linux. A broad range of program analysis tools, called *Pintools*, can be built under this framework. The instrumentation of binary at runtime dynamically generates code, allows generic morphing, and alleviates the need to modify or recompile the source.

Here, we designed the GC Pintool in Fig. 2 to identify the dereferenced memory addresses allocated during the application execution, but never freed up. Then, after the end of each block execution all dereferenced memory can be released using this tool and logged. PIN can be executed in the probing mode for dynamic invocation of GC Pintool as and when required.

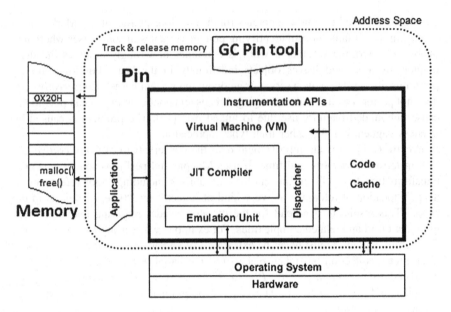

Fig. 2 Software architecture of PIN

4 Implementation

In this section, we will discuss the implementation infrastructure for GC Pintool. The GC Pintool follows the basic principle of garbage collection, that is

1. Find inaccessible objects in memory and
2. Free these objects and reclaim the memory.

Further, since the GC Pintool is a dynamic instrumentation tool, so every action happens at runtime which ensures zero modification in the source code. Now, first the modus operandi of the tool is explained and then later various Pin APIs that have been used by the Pintool to make GC possible at run time are described. Throughout the discussion carried over in this section, the two terms namely Mutator: to symbolize the application program; and Collector: to signify the GC Pintool, has been frequently used. Initially, when mutator starts its execution under Pins control, the collector sets its breakpoint as per the instrumentation routines. The major breakpoints for the instrumentation are as follows:

1. Calling of any user function
2. Calling of any Memory Allocator function like malloc(), calloc(), and realloc().
3. Calling of any Memory Deallocator function like free().
4. Return from any user function

Whenever any function of mutator is called, it is recorded by the collector. Collector further waits for any heap allocation that is, dynamic memory allocation to be made by the mutator. The collector captures the memory location of the heap being allocated and save it in its master data structure along with some other information in which scope (function) the allocation is made. In this manner, all the allocations made by the mutator get captured by the collector in its data structure. If there is any deallocation made by mutator, it is also checked in the data structure and that log is removed from the data structure. When any function completes its execution, the collector program sweeps away all dynamic memory allocated by that function. Important is, collector takes care of the allocations made against global or, if any reference is passed to some other function then in those cases, collector does not make any deallocation rather that the scope is changed for those allocations suitably. This way the collector ensures that there is no illegal or premature deallocation. Now, below we will present the GC Pintool algorithm

```
Algorithm: GC Pintool (GCMAP<fname, memaddr>, fname, memaddr)

  This Algorithm Logs the details of Memory allocated dynamically and Free the memory
  when it is Leaked or becomes Garbage, simultaneously during the execution of C Program.

// Global Initialization:
GCMAP<fname, memaddr> ; // Global Map to keep Garbage Record
char* fname; // To store Name of function being called
Address memaddr, mem_free; // To store the memory addresses returned by Allocator function

// This part will execute when any user/main function is called.
fname = Function_Name;

IF (Memory Write)
  IF(Memory Write == Global Reference)
    Global; fname = main; // scope change
    GCMAP.insert(<fname, memaddr>);
  End IF
  IF(Memory Write == Parameter Reference)
    fname = Function_Name; // scope change
    GCMAP.insert(<fname, memaddr>);
  End IF
// This part will execute when Mem. Allocator functions
// that is, malloc(), calloc() & realloc() are called.
GCMAP.insert (<fname, memaddr>);
// memaddr store the address returned by  memory allocator function

// This part will execute when free() function is called
mem_free = Address returned by free();
While (GCMAP.begin() to GCMAP.end())
    IF(GCMAP.memaddr == mem_free)
      GCMAP.erase (<memaddr>);
    End IF
End While

// This part will execute at the Exit of any user/main function.
While ( GCMAP.begin() to GCMAP.end() );
  free(GCMAP.memaddr) // Garbage Created by Returned Function
GCMAP.erase(GCMAP.memaddr);
```

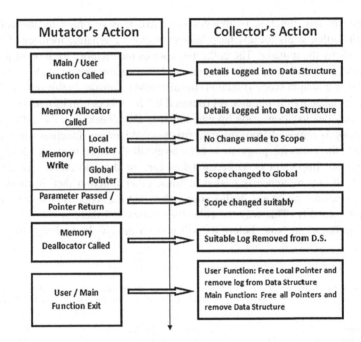

Fig. 3 GC approach using dynamic instrumentation

The Flow Chart depicting the GC approach as described in above algorithm is as shown in Fig. 3.

In order to keep the logs of memory addresses which are allocated or deallocated dynamically, a master data structure has been used by the GC Pintool. During the execution of the program, this data structure keeps on changing as per the instrumentation and accordingly GC Pintool collects information regarding performing the garbage collection at a suitable point of time. Figure 4 shows the typical transition in the data structure during the program execution.

Some important Pin APIs used in the Pintool are discussed below

RTN_InsertCall(RTN Rtn, IPOINT Action, AFUNPTR Funptr, ⋯) This API is used to insert a call relative to a routine (rtn) and a suitable action is taken like IPOINT_BEFORE to call funptr before execution of rtn, or IPOINT_AFTER for immediately before the return from rtn. There are various IARG_TYPE arguments to pass to funptr.

PIN_CallApplicationFunction(Const CONTEXT * Ctxt, THREADID Tid, CALLINGSTD_TYPE Cstype, AFUNPTR OrigFunPtr, ⋯) This API allows the tool to call a function inside the application. The function is executed under control of Pin's JIT compiler, and the application code is instrumented normally. Tools should not make direct calls to application functions when Pin is in JIT mode.

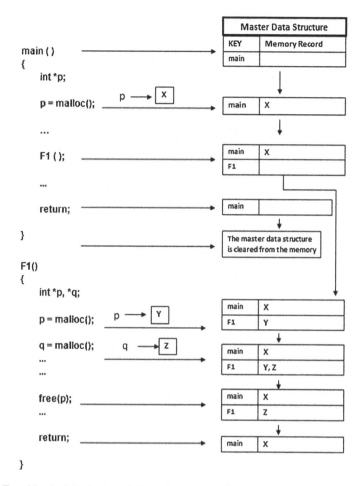

Fig. 4 Transition in data structure during program execution

For that reason, to call mutator's free() this API has been used. This API in turn deallocates the memory that has been allocated dynamically by the mutator.

PIN_SafeCopy(VOID * Dst, Const VOID * Src, Size_T Size) This function is used by our tool to ensure safe access to the original content of the application's memory from our tool. This API is helpful in confirming the type of reference, i.e., local, global, etc. On this basis, the GC Pintool makes decision regarding the scope of the allocation.

The instrumentation algorithm in GC Pintool has been tested on various benchmark programs and is found to perform accurate detection of memory errors and garbage collection for C programs.

5   Functional and Performance Testing

In this section, we first present the test plan for correctness of GC Pintool and then discuss how the performance of the tool has been assessed.

5.1   Correctness of GC Pintool

Table 1 presents a test plan for the GC Pintool. It covers different scenarios for memory issues in terms of variables of a C program while dynamic memory allocation is used. Based on the test plan, a benchmark test suite comprising 168 small to medium C programs was created. In addition, 10 C programs with memory issues as reported in different user forums were also added to the test suite. GC Pintool performed correctly for the whole test suite.

5.2   Performance of GC Pintool

Since GC Pintool relies on dynamic instrumentation, the performance of a C program running under the tool is expected to be significantly degraded compared to a natively running C program. To estimate the performance impact, we recall that GC Pintool performs two primary tasks: (1) Detect memory leaks and other issues and (2) Releases memory that is no more usable.

The detection is implicit because unlike usual memory tools, no report is made to the user; rather the information is used by the tool to release appropriate resources. So, in this part, GC Pintool closely resembles the functionality of Valgrind and we can make a direct comparison. It may be noted that Valgrind experiences 10–50 times slowdown[1] and we would expect similar behavior.

In comparison with the Valgrind's behavior, we prepared eight C programs containing a variety of memory scenarios from Table 1 and executed them for five varying data sets (using 10 MB–2.5 GB memories) under both Valgrind and GC Pintool.[2] While GC Pintool always performed correctly, we find that it runs about 35 % faster compared to Valgrind. This is rather encouraging, given that Valgrind is a widely used tool.

In the other part (where memory is actually released to achieve GC), there is no reference to compare against. So, we run GC Pintool (with doing GC) and compare against the run of the tool that only performs the detection, but does not release. We find that for the above programs, our tool runs about one order slower when it performs GC.

---

[1] "2.1. What Valgrind does with your program" in http://valgrind.org/docs/manual/manual-core. html.

[2] We needed to tweak the tool to report memory issues.

Table 1 Test plan for GC pintool

Type	Func.	Block	Case 1	Case 2	Case 3	Case 4
Scenarios for local variables						
Local	Single	Function	1 Alloc.	2 Alloc.	2 Alloc.	2 Alloc.
			No Dealloc.	1 Dealloc.	2 Dealloc.	No Dealloc.
		Nested	Alloc. outer	Alloc. outer	Alloc. inner	Alloc. inner
			Dealloc. outer	Dealloc inner	Dealloc. outer	Dealloc. inner
		Loop	Alloc.	Alloc. outer	Alloc. inner	Alloc. inner
			No Dealloc.	Dealloc. outer	Dealloc. outer	Dealloc. inner
		Recursive	Alloc.			
			No Dealloc.			
	Two	Function	1 Alloc.	2 Alloc.	2 Alloc.	2 Alloc.
			No Dealloc.	1 Dealloc.	2 Dealloc.	No Dealloc.
		Nested	Alloc. outer	Alloc. outer	Alloc. inner	Alloc. inner
			Dealloc outer	Dealloc. inner	Dealloc. outer	Dealloc. inner
		Loop	Alloc. outer	Alloc. inner	Alloc. inner	
			Dealloc. outer	Dealloc. outer	Dealloc. inner	
		Recursive	Alloc.			
			No Dealloc.			
Scenarios for global variables						
Global	Single	Function	1 Alloc.	2 Alloc.	2 Alloc.	2 Alloc.
			No DeAlloc.	1 Dealloc.	2 Dealloc.	No Dealloc.
		Nested	Alloc. outer	Alloc. outer	Alloc. inner	Alloc. inner
			Dealloc. outer	Dealloc. inner	Dealloc. outer	Dealloc. inner
		Loop	Alloc outer	Alloc inner	Alloc inner	
			Dealloc outer	Dealloc outer	Dealloc inner	
		Recursive	Alloc.			
			No Dealloc.			
	Two	Function	Alloc in 1	2 Alloc in 1	2 Alloc in 1	2 Alloc in 2
			Dealloc in 2	No Dealloc	1 Dealloc	No Dealloc.
Scenarios for parameters						
Param	Two	Function	One Alloc., Reference returned back	Two Alloc., One Dealloc, One Reference returned back	Two Alloc. Two Dealloc.	Reference is Pointer to Pointer

Type	Func.	Block	Cases			
Scenarios for member variables and special cases						
Data member	Single	Function	Allocation against a data member of structure			
			Allocation against a pointer to a structure			
			Allocation against pointer to a structure till out of memory			
Local			Pointer reassignment			
			Allocation against the array of pointers			

6 Conclusion

Memory Management is an important aspect for any programming language. Inefficient management of the memory in a program may lead to various consequences like memory leak, dangling pointer, double free error, etc., resulting in slow down of the program, fragmentation, incorrect execution, premature GC, or program termination. As a solution to these memory management problems, specifically for C programs, a novel garbage collection approach has been proposed and developed in this paper that uses dynamic binary instrumentation which is accurate and does not need any modification in the source code.

The GC Pintool has been successfully tested over a large set of distinct test codes. It takes care of all the dynamically allocated memory whether it is allocated against a local pointer variable, global pointer variable, or passed as parameter to some other function. The tool deallocates the reserved memory at proper time during the execution of the program. The tool has been shown to successfully detect any kind of memory leak error and performs the garbage collection suitably. Further, it is capable of handling issues like pointer reassignment, allocation made against the data members of an object, arrays, array of pointer, and pointer to an array. GC Pintool overcomes the drawbacks of *Conservative GC* [10, 12] and *Precise GC* [11] and has been shown to run faster than Valgrind for memory leak/error detection.

The GC Pintool is a work in progress. The present version is a functional prototype, intended to operate on moderately large C programs to provide an understanding of its behavior and to provide a platform for adding future enhancements. The present tool may be extended in the following directions:

- *Support for C++* Internally GC Pintool already has most of the infrastructure required for C++. We can log the memory allocated and deallocated by new and delete operators, but the challenge will be to deal with the constructors and destructors of an object.
- *Improving Efficiency* At present, GC Pintool performs selective instrumentation which has helped to restrict the slowdown while dynamic instrumentation is used. This is manifest in GC Pintool which has a better performance than Valgrind when it is used only for detection of memory issues, but no *Garbage Collection* (GC) is actually done.

  Unfortunately, while doing the GC, GC Pintool is confronted with a second level of slowdown because it needs to dynamically call appropriate free function for any memory that is about to be leaked. It may be noted that this call actually does not exist in the user code and hence needs to be inserted at the runtime. GC Pintool achieves this by using PIN_CallApplicationFunction, a function of PIN. Incidentally, this function has a lot of overhead and substantially slows down the GC Pintool further. We are working on a few schemes to overcome this shortcoming—one is to move to a lazy collection strategy and the other is to use the fact that GC only needs to call a fixed function, namely, free.

Further, the multimap structure used to log memory allocations and dealloca-tions, may be improved to reduce space complexity.

- *Testing on Legacy programs* GC Pintool should be tested on large and complex legacy programs so as to ensure its applicability in production environments.

References

1. Cohen, J.: Garbage collection of linked data structures. Comput. Surv. **13**(3), 341–367 (1981)
2. Cohen, J., Nicolau, Alexandru: Comparison of compacting algorithms for garbage collection. ACM Trans. Program. Lang. Syst. **5**(4), 532–553 (1983)
3. Wilson, P., Johnstone, M., Neely, M., Boles, D.: Dynamic storage allocation: a survey and critical review. In: Proceedings of the International Workshop on Memory Management, Kinross Scotland (UK) (1995)
4. Jones, R.E.: Garbage collection: algorithms for automatic dynamic memory management. Wiley, Chichester (1996)
5. Harold McBeth, J.: On the reference counter method. Commun. ACM **6**(9), 575 (1963)
6. McCarthy, John: Recursive functions of symbolic expressions and their computation by machine. Commun. ACM **3**, 184–195 (1960)
7. Ungar, D.: Generation scavenging: a non-disruptive high performance storage reclamation algorithm. In: Proceedings of the ACM Symposium on Practical Software Development Environments, pp. 157–167 (1984)
8. Blackburn, S.M., Cheng, P., McKinley, K.-S.: Myths and reality: the performance impact of garbage collection. In: Sigmetrics—Performance 2004, Joint International Conference on Measurement and Modeling of Computer Systems, New York (2004)
9. Dijkstra, E.W., Lamport, L., Martin, A.J., Scholten, C.S., Steffens, E.F.M:. On-the-fly garbage collection: an exercise in cooperation. Commun. ACM **21**(11), 965–975 (1978)
10. Boehm, H.-J.: Space efficient conservative garbage collection. In: Proceedings of the ACM SIGPLAN 1993 Conference on Programming Language Design and Implementation, pp. 197–206. ACM Press (1993)
11. Rafkind, J., Wick, A., Regehr, J., Flatt, M.: Precise garbage collection for C. In: Proceedings of ISMM 09 International Symposium on Memory Management, pp. 39–48. ACM Press (2009)
12. Boehm, H.-J.: Bounding space usage of conservative garbage collectors. In: Proceedings of the 29th ACM SIGPLAN-SIGACT Symposium on Principles of Programming Languages, pp. 93–100. ACM Press (2002)
13. Hirzel, M., Diwan, A., Henkel, J.: On the usefulness of type and liveness accuracy for garbage collection and leak detection. ACM Trans. Program. Lang. Syst. **24**(6), 593–624 (2002)
14. Luk, C., Cohn, R., Muth, R., Patil, H., Klauser, A., Lowney, G., Wallace, S., Reddi, V.J., Pin, H.K.: Building customized program analysis tools with dynamic instrumentation. In: Proceedings of the 2005 ACM SIGPLAN Conference on Programming Language Design and Implementation (Chicago, IL, USA, June 12–15, 2005)

A Study on Software Risk Management Strategies and Mapping with SDLC

Bibhash Roy, Ranjan Dasgupta and Nabendu Chaki

Abstract In recent years, despite several risk management models proposed by different researchers, software projects still have a high degree of failures. Improper risk assessment during software development was the major reason behind these unsuccessful projects as risk analysis was done on overall projects. This work attempts in identifying key risk factors and risk types for each of the development phases of SDLC, which would help in identifying the risks at a much early stage of development.

Keywords Risk management · Risk models · SDLC · Technical risk

1 Introduction

Software project management is crucial for development, services and maintenance of software products. Management of diverse activities during software engineering process needs to be handled carefully for any software project. One of the most important yet often overlooked aspects in the complete process is risk and its management [1]. A risk may be considered as a probabilistic term that has the potential to affect the overall project in a negative way. Failure of the projects, especially IT projects, is often due to these unwanted and rather less explored

B. Roy (✉)
Tripura Institute of Technology, Tripura, India
e-mail: bibhashroy10@yahoo.co.in

R. Dasgupta
National Institute of Technical Teachers' Training and Research, Kolkata, India
e-mail: ranjandasgupta@ieee.org

N. Chaki
University of Calcutta, Kolkata, India
e-mail: nabendu@ieee.org

© Springer India 2016
R. Chaki et al. (eds.), *Advanced Computing and Systems for Security*,
Advances in Intelligent Systems and Computing 396,
DOI 10.1007/978-81-322-2653-6_9

threats or issues [2, 3]. Some works during the last two decades on risk management strategies have emerged as survival component for software projects [4, 5]. However, it is important to identify the possible risks in all the stages of the software development process so that proper mitigation strategy can be adopted at the appropriate level to reduce the possible financial and temporal loss.

Any software project may face different categories of risks, either internal or external, in its engineering process. There are different kinds of risks encountered in different phases of software development life cycle (SDLC) and these risks can be classified as technical risks and nontechnical risks. The internal risks are due to the factors within the organization and external risks come from outside the organization and are difficult to control. Software risks can be grouped into project risks, process risks, and product risks. Risks for software project management have been classified as Known risk, Unknown risk, Predictable risk, Unpredictable risk [3]. Another approach used in [6] identifies risks as Technical risk, Management risk, Financial risk, Contractual and legal risk, and Personal risk. On the other hand in [7], it was observed that risks are identified as Business risk, Commercial risk, Economical risk, Project risk, Product risk.

In the last few decades the researchers all over the globe had done lots of works and various risk models had been proposed and applied at different cases with various levels of success. All such models have lots of merits; however, no single model can be applied to all cases. Moreover, for a complex and big projects looking the risk analysis from the top level might not be appropriate as the risks occurred in the lower level might not be identifiable if the tool for identification is not applied at that level where the actual risk might occur. In this paper, risk management models are reviewed in such a way that the models can be mapped with the different stages of the life cycle of a software development process. For this purpose, major objectives and capacities of various risk management models had been studied (Sect. 2). All steps of activities of a software development process had also been studied with an eye to identify the types of risks that may occur at each level (Sect. 3). In Sect. 4, mapping has been done from the outcomes of Sects. 2 and 3 so that insight knowledge can be extracted from the final result to design the mitigation strategy.

2 Risk Management Models

Risk management is a set of activities required to manage risk. Organizations who apply risk management methods and techniques have greater control over the projects [8]. The risk management methods that are being proposed till date follow mainly two characteristics—probability and impact. Most of the proposed methods are static in nature that performs qualitative and quantitative analysis to assess and control the risks. Most of the methods divide the risk management into some basic processes like *risk identification, risk analysis, risk planning or mitigation, risk monitoring and control,* etc. [3, 9]. It is very much essential to identify the causes as well as effects of the risks for a comprehensive analysis of risks and many of the methods perform this but concentrate on a single risk.

One of the oldest as well as mostly discussed risk management models is Barry Boehm's (1991) BOEHM Model [1–3, 10] that emphasizes on the concept of 'risk exposure' and can be applied to almost any software-related project. BOEHM divides the risk management into two primary steps—*Risk assessments (risk identification, risk analysis, and risk prioritization) and risk control (risk management planning, risk resolution, and risk monitoring)*. Boehm proposed an effective risk control strategy where each risk is thoroughly tracked with the aim that this risk either can be eliminated or its effect into the project in terms of cost or time can be reduced to certain extent. Boehm has discussed different tools like checklists, cost models, cost–benefit analysis, etc., in each step, however, no standard project metrics was used as a tool for risk management. This method can be applied in all the phases of SDLC as risk analyzer. However, new risks that are occurred during the development process were not taken into consideration.

Software Engineering Institute in 1993 proposed SEI-SRE [2, 3, 11, 12] (Software Engineering Institute, Software Risk Evaluation) model that provides a framework for risk evaluation to overcome project failures. This model concentrates on practical aspects of a project that makes it eligible to be applied successfully in IT projects as a decision-making tool. However, there is no scope for modification in the template-based design and sometimes inconclusive outcomes might occur as it mostly relies on the experiences of project personnel. Unlike Boehm, it has proper method to measure and evaluate the effectiveness of risk treatment; however, fails to perform review of risk information periodically as performed by Boehm model.

Ronald P Higuera et al. in 1994 proposed Team Risk Management (TRM) Process Set [5, 12, 13] that emphasizes on team structures and its activities for managing risk in each phase of SDLC involving all individual connected to the project to ensure continuous risk management throughout the project development. This approach follows continuous risk evaluation process where new risks are considered and mitigated or resolved risks are automatically removed from the threat list and thereafter updated risk status is communicated to all the individuals connected in the project. Like TRM, Agle et al. (2003) also proposed [14, 15] a risk handling mechanism related to team structure in multi-team environment.

Another widely recognized method developed by Jyrki Kontio et al. in 1996 is known as RISKIT [2, 3, 12, 16]. This risk management cycle performs *risk identification, risk analysis, risk monitoring,* and performs qualitative analysis to prioritize according to probability and impact. It uses a graphical method called Risk Analysis Graph (RAG) to monitor risk scenario development and the concept of utility loss is used to assess the impact of risk. This model may be applied mainly in large organizations, IT projects, and Nokia Telecommunications is one of its major users. However, it is very much flexible to be implemented in different categories of projects or areas such as business planning, marketing, and technology-related fields. Due to the absence of correlation between risk estimation and risk metrics, this method suffers inaccurate prediction possibilities of potential risks.

In the year 1997 two different models were proposed: Software Engineering Risk Understanding and Management (SERUM) [2, 3, 17] by D. Greer and Risk Maturity Model (RMM) [18] by David A. Hillson where SERUM performs both *implicit risk and explicit risk* management and RMM measures the maturity level of an organization. Though SERUM is typically designed for software projects, it suffers time management problem in analyzing risks. Unlike other maturity model such as CMM, EFQM, etc., RMM is completely based on risk management. It performs different risk management approaches of an organization to assess the maturity level in managing risks.

Software Risk Assessment Model (SRAM) [19] and SoftRisk [5, 12, 20] are the two different risk management approaches that came out during the year 2000 where SRAM uses a set of comprehensive questionnaire for different critical risk elements to prioritize risks and SoftRisk generates a graphical tool helping the managers to perform risk control mechanism. Like TRM model, SoftRisk approach also considers the new risks in its continuous risk management process.

In 2007, William G. Snekir et al. proposed Enterprise Risk Management (ERM) method that performs risk assessment with the help of graphical decision tree and quantitative analysis.

Armestrome and Adens (2008) [15, 21] gave a very decent idea of identifying most prominent area of exposure to risk to characterize those areas into different risk factors and to prioritize them to perform risk management strategies. On the contrary, Software Risk Assessment and Estimation Model (SRAEM) [3, 4, 22] having weak risk identification strategies estimates the sources of risks from different paradigms and compute prioritization and ranking by MCRSRM (Mission Critical Requirements Stability Risk Metrics) followed by quantitative assessment. Unlike SRAEM, Analyzer for Module Operational Risk (ARMOR) [23] automatically identifies the operational risks of software program modules. Being a module-based approach this model can perform risk management in every stage of software development and like SoftRisk model, can perform various types of data analysis. A modified approach of SRAEM is Software Risk Assessment and Evaluation Process using model-based approach (SRAEP) [3, 4, 24] that aims at risk assessment and risk prioritization. This model uses SFTA (software fault tree approach) to identify and analyzes the risk and uses RRL (risk reduction leverage) for risk measurement.

Hoodat and Rashidi (2009) [15, 25] have proposed different classifications of risk based on indexing of different risk factors followed by calculating their impact on software project. Different categories of risk identified by them are (a) *Internal risk and external risk.* (b) *Process risk, product risk, and project risk.* (c) *Performance risk, cost risk, and scheduling risk.* (d) *Requirement risk, cost risk, scheduling risk, quality risk, and business risk.* A similar risk management activity performed by Software Project Risk Management model (SPRMQ) (2011) [7] requires experience of project manager to manage software product risk. However, unlike other models it does not consider the external risk. Another approach as proposed by DANNY in 2006 [15, 26] reduces operational risk by performing risk classification and their qualitative analysis with an aim to save resources with a consideration for small-sized projects or available funded projects.

Suebkuna and Ramingwong (2011) refined the approach of Project Oriented Risk Management Model (PRORISK) [27] with a proposal to link project management and risk management having two phases of risk management: risk assessment and risk control. It uses a risk database to record risk control related information such as its impact, probability, type, mitigation strategy, etc.

Shahzad et al. (2011) proposed risk "handling and avoidance mechanism" in Risk Identification, Management and Avoidance Model (RIMAM) [3, 28]. This model provides a stepwise procedure of risk handling methodology that enables the development team to handle risks locally. This model is suitable to be applied in small and medium scale of software with tight budgetary and short-time period.

Project Risk Network Model (PRM) [6, 12] as proposed by Linda Westfall (2011) is one of the well-known methods detailing a Decision Support System framework consisting of five major steps: risk identification, risk assessment, risk analysis, risk response planning, and risk monitoring and control enabling project managers in choosing a set of risk mitigation action with minimum loss. The drawback of this model is that it uses the classical method for risk identification and evaluation during its initial phases with an assumption that risks are independent, whereas there are lot of interactions and influences between different risks in the project development.

Software Engineering Risk Index Management (SERIM) (2014) [2, 3, 29] is one of such risk management models that is typically designed for software project development focusing mainly on the assessment of risk factors (*Organization, estimation, development, methodologies, tools, risk, culture, usability, correctness, reliability, and personnel*) with periodic measurements on high priority risk areas throughout software development stages. Though this method lags in providing explicit guidelines to identify risks that may involve in the project, medium-sized organization may consider it instead of expensive methods.

Most recently Loutchkina et al. (2014) proposed System Integration Technical Risks' Assessment Model (SITRAM) [12] that integrated Bayesian Belief Networks (BBN) and Parametric models (PM) providing statistical information for improving risk management during large software development.

In this section, we presented observations on some of the well-known risk management models. These models cover really an wide and diverse range of objective right from identification of risks to risk mitigation, risk assessment, controlling of risks, prioritization of risks, risk probability computation, and even proposing a four level of risk maturity model. Some of the existing risk models are good specifically for either small or large organizations. On the other hand there exist risk models that especially deals with risks associated with software projects only. Many of these have been implemented using suitable tools. However, in the current state scenario only a few models exist that considers risks associated with different stages of SDLC. The study further reveals that such models, in use, lack

Table 1 Observations of different risk models

SN	Methods/models/proposed	Observations
1	BOEHM [10]	Does not handle generic risk; works on risk analysis paradigm principle
2	SEI-SRE [11]	Generates a template-based design that results in inconclusive outcomes due to less scope for modification
3	RISKIT [16]	Does not collaborate risk estimation and risk metrics, thus reducing the prediction possibilities of potential risks
4	SERUM [17]	As it performs a continuous evaluation of risks, hence time management holds the key role as risk element in the project
5	SERIM [29]	Good for small organizations; handles multiple projects for analyzing software risks; lacks explicit guidelines on using information to identify possible risks in the project
6	SRAM [19]	Risk ranking is done by AHP and entropy method. It does not handle marketing risk
7	Agle et al. [14]	Handles team structure; does not consider funding and resources
8	Danny [26]	Performs classification of risk by quantitative analysis; aims at saving resources
9	Armestrong [21]	Identifies the risk exposure areas and prioritizes them in respect to business context
10	Rashidi [25]	Perform risk classification and risk indexing
11	SRAEM [22]	Risk prioritization and ranking is computed by MCRSRM, decision through quantitative assessment; model focuses on external risks related to the requirement analysis
12	SRAEP [24]	Model uses SFTA to identify and analyses the risk and RRL for risk measurement; follows models based approach
13	SPRMQ [7]	Well suited for handling the product risk; does not consider external risks such as marketing risk, organizational risk, etc.; uses avoidance, minimization, and contingencies strategies
14	RIMAM [28]	Works on the principle of "*handling and avoidance mechanism*"; some of the risks can be handled locally
15	TRM [13]	Follows all the steps of SEI; handles new risks and risk status are communicated to all individuals
16	SoftRisk [20]	Documents all types of risks; performs qualitative and quantitative analyses; Consider new risks in an iterative process
17	ARMOR [23]	Identifies source of risk and suggests solution to reduce risk levels; uses regression analysis to validate generated risk model
18	RAT [41]	Performs hybrid assessment of risks in five phases; risks are ranked based on ranking matrix

(continued)

Table 1 (continued)

SN	Methods/models/proposed	Observations
19	ERM [42]	Evaluates level of an organization to propose risk assessment tool using graphical decision trees and quantitative analysis
20	PRORISK [27]	Links project and risk management toward developing a risk database; handles six types of risks for software projects
21	RMM [18]	Provides the benchmark to an organization to assess its maturity level in terms of project risk management
22	PRM [6]	Works on the assumption that risks are independent which may lead to incorrect risk assessment
23	SITRAM [12]	Provides statistical information in decision-making of risk management using Bayesian belief networks and Parametric models

completeness from one or more perspectives. Based on the above discussion, we have summarized the key features of these models in Table 1. However, other research works [15, 30–40] deal with various special issues like Risk management in requirement engineering, Risk-based testing, Project Risk dependencies, etc., and not included in Table 1.

Based on above observations of risk management models under discussion it can be summarized as indicated in Table 2 that different models are strong enough to be applied in different application domains.

Risk management models that are discussed in this section involve different participants in its risk management process and are summarized in Table 3.

Among the risk management models considered in our discussion, only few models emphasized on the importance of project size in its risk management strategies. It is general consideration that if a risk management model can handle a large-sized project then it will surely be able to handle medium-and small-sized project too. The medium- and small-sized projects face mainly two constraints: very tight schedule and a limited budget [34]. These constraints do not permit to go for a risk management strategy that may incur more costs and time [15]. Rather than going for those complex and costly risk management models, a simple and effective model would be used for relatively smaller projects. It has been found that there are monitory losses in medium and smaller projects due to improper handling or not considering the risks related to the projects [34]. During last decade the concentration has been given on the risk management strategies for small and medium projects [2, 3, 5]. Different researchers have proposed risk models for small and medium software projects [15]; however, proper and detailed description of management process would require some standards. Risk management models under discussion may be compared on the basis of effectiveness on budget, time, and size of the project and is tabulated in Table 4.

Table 2 Risk models categorization based on their application domain

SN	Methods/models/proposed by	Application domain
1	BOEHM [10], SEI-SRE [11], SPRMQ [7], PRORISK [27], ARMOR [23], RISKIT [16]	Software/IT projects
2	TRM [13], Hoodat_Rashidi [25]	Can be used as team risk management tool in software development process
3	SoftRisk [20], PRM [6]	Flexible to be implemented in different categories of projects of any domain
4	RMM [18]	Can be used by any organization to assess its risk maturity level
5	ERM [42]	To assess an organization's level
6	SERIM [29], SRAM [19], RIMAM [28], SRAEM [22], SRAEP [24]	Small- and medium-sized software projects
7	Danny [26]	Small projects with no costs risk involvement
8	Armestrong [21]	Projects with budget constraints
9	RAT [41]	Web application
10	RISKIT [16], SERUM [17]	Existing software's new version release
11	SITRAM [12]	Large and complex software projects

Table 3 Risk models categorization based on participants

SN	Methods/models/proposed by	Participants in the process
1	BOEHM [10], RISKIT [16], SRAEM [22], SRAEP [24], SoftRisk [20], ARMOR [23]	Risk management team
2	PRM [6]	Risk management team, project manager
3	RIMAM [28]	Risk management team, development team
4	RMM [18], SRAM [19]	Risk assessment team
5	Armestrong [21], Hoodat_Rashidi [25]	Project managers
6	SERIM [29], SPRMQ [7], PRORISK [27], Danny [26]	Project managers, risk managers
7	ERM [42], SEI-SRE [11]	(Experienced) risk managers
8	RAT [41]	Risk managers, project development team
9	SERUM [17]	Software development team
10	TRM [13]	Development team, stakeholders, customers, users
11	SITRAM [12]	Risk assessment team

Table 4 Risk models categorization based on projects size, cost, and application

SN	Methods/models/proposed by	Effectiveness		Applicable to projects size
		Budget	Time	
1	BOEHM [10], RISKIT [16], SERUM [17], PRM [6], RMM [18], Hoodat_Rashidi [25], SoftRisk [20], ARMOR [23]	Tight	Scheduled	Large
2	SEI-SRE [11], SPRMQ [7], Armestrong [21]	Tight	Scheduled	Large, medium
3	ERM [42]	Tight	Scheduled	Large, medium, small
4	SERIM [29], SRAM [19]	Tight	Scheduled	Small, medium
5	RIMAM [28]	Tight	Critical and short	Small, medium
6	SRAEM [22], SRAEP [24]	Tight	Critical	Small, medium
7	TRM [13], RAT [41], PRORISK [27]	Tight	Scheduled	Small, medium
8	Danny [26]	Available	Scheduled	Small
9	SITRAM [12]	Available	Scheduled	Large and complex

3   Risks at Each Level of SDLC

Software project development is a systematic process and risks may occur in every stage of this process. Thus it is essential to look the entire risk management mechanism from a different angle. Conventional approaches of study for internal risk, external risk, known risk, unknown risk, etc., are necessary. However, these cannot be considered as sufficient to identify the chances of occurrence, impact of such risk(s) in terms of cost and time, etc., at each level of SDLC. As for example, if a *risk of conflicting requirements* is left unidentified at the requirement specification level as no conflict identification technique (mechanism) has been applied at that level, the risk will naturally be carried forward and in worst case it might been slipped as undetected and carried up to implementation. The risk, when occurred might not only cause some serious business loss, identification, and mitigation of the problem would also be very cumbersome and extremely difficult. On the other hand, if appropriate mechanism be applied at every stages of SDLC, it not only will identify and arrest a lot of risks to propagate further and cause serious damages, maintenance overhead will also be minimized and more robust software can be offered to the users. Moreover, the impact of the risk might be many folds higher and serious, if it is propagated forward and occurred at a later stage, particularly, when the customer's dependency on the product has become higher and insepa- rable. Loss of confidence, if happens, on the product and services is another issue which neither can be measured and sometime may lead to irreparable damage and cause change in business sentiments.

The necessity of new risk categorization at a more micro level is thus a necessary requirement where risks can be categorized as per their chances of occurrences in the various development stages. As risk management strategy involves both risk assessment and risk mitigation, hence this classification will help in stagewise risk identification and mitigation during software project development. Software project risks may be grouped as per their occurrences in the phases of SDLC, i.e., requirements and planning, designing, coding, application and maintenance, and other related parameters like scheduling, cost, quality, and business are also affected.

As per SEI [1, 4, 5, 11] the following risk factors may be associated to software development.

1. Incorrect resources estimation
2. User/customer uncertainty
3. Ambiguous requirements
4. Improper design risk
5. Development system and risk with development system
6. Inadequate management process
7. Improper work environment

Unlike other generic project risks, technical risks are matter of concern for software and IT projects and generally lead to failure of functionality and performance [12]. SEI also identifies following technical risks associated with technology related projects

- Lack of strategic framework or conflict over strategy
- Lack of adaptation to technological change
- Supplier/vendor problems
- Poor management of change
- Too much faith in ability of the technology to fix the problems

In the context of SDLC, the various causes or factors have been identified (Tables 5, 6, 7, 8 and 9) as technical risks at different phases against some key factors.

4 Mapping of Models for Different Stages

Different risk management models discussed so far were performed considering overall scenario of the project. Whereas risks in software projects may occur in any of the phases of SDLC and depends on separate strategies adopted for individual phases of SDLC. Software risks have been broadly grouped in literature [6] as nontechnical risk (project and business) or as technical risk. However, risks related to software project development also depend on factors like product engineering, development environment, and program constraint. Again, there are scheduling risks and quality risks that come under planning phase. In addition to software

Table 5 Technical risks at requirements analysis and planning phase

Key factors	Overview of key factors	Technical risks
Stability	Risks due to instability in requirement	Continuous changing requirements
Completeness	Incomplete or unrealistic planning	Inaccurate sizing of deliverables
		Unrealistic time schedule for individual module development
Clarity	Improper requirement or inadequate analysis	Improper definition of requirements
		Inadequate software project risk analysis
Validity	Less or no knowledge about validity of existing tool, inaccurate estimation	Inaccurate cost estimation
		Inaccurate quality estimating
		Inadequate software policies and standards
		Less knowledge about the availability of resources
		Incorrect estimation of resources
		Inability to estimate the scope for reusability of existing modules
Feasibility	Lack of documentation, reusable resources	Nonavailability of documentation of previous projects
		Lack of reusable requirements
		Lack of reusable documentation
Scalability	Scalability of resource requirements	Inaccurate metrics
		Inadequate assessments

Table 6 Technical risks at design phase

Key factors	Overview of key factor	Technical risks
Functionality	Incorrect function design	Inexperienced software module designer
Difficulty	Inability to design or unavailability reusable design	Inadequate tools and methods for quality assurance
		Inadequate tools and methods for software engineer
		Lack of reusable data
		Lack of reusable design
		Lack of reusable architecture
Interfaces	Less knowledge of interface design	Incorrect interface design
Performance	Incomplete designed module	Error-prone module designed
Testability	Unrealistic design	Inexperienced software module designer
Hardware	Little or no knowledge of hardware constraints	Incorrect hardware simulation or interface

Table 7 Technical risks at coding phase

Key factors	Overview of key factor	Technical risks
Functionality	Non-functioning of development team	Inexperienced technical staffs
		Inexperienced development team
		Malpractices of technical staff
Feasibility	Non-mapping between design and coding	Unrealistic module designed to develop using existing technology
		Difficult project modules integration
Testing	Inaccurate or unrealistic development	Wrong integration of modules developed
Availability	Nonavailability of technology, reusable resources, and experience	Unavailability of advanced technology
		The existing technology is in initial stages
		Less reusable code available in the organization
		Inadequate technical training to the staff
		Lack of reusable human interfaces
		Lack of any specialization
		Poor technology investment
		Slow technology transfer
Coding/implementation	Excessive coding and development	Product is complex to implement using existing technology and technical staffs
		Excessive load to development team
		Overschedule due to erroneous coding
		Excessive coding due to lack of standard programming guidelines

Table 8 Technical risks at testing phase

Key factors	Overview of key factor	Technical risks
Environment	Inadequate testing environment	Unable to identify ambiguity in the developed product Lack of reusable test plans, test cases, and test data
Validity	Inexperienced testing team	Inexperienced testing team Inability to identify and to fix problems or errors
Product	Incorrect testing on product	Improper testing strategies
System	Improper developed system	Improper defined objectives to the testing team

Table 9 Technical risks at application and maintenance phase

Key factors	Overview of key factor	Technical risks
Maintainability	Non-flexible, platform-dependent design, high cost	No or less scope for change in the system Platform-dependent application High maintenance costs
Reliability	Non-reliable, non-satisfaction developed system	Unsatisfied customer of the developed product Partially developed product Low productivity
		Low user satisfaction
Safety	Inadequate documentation, tools for maintenance	Overbudget due to corrective maintenance
		Lack of proper help or manuals to the users
		Inadequate tools and methods for technical document
Security	Less or no security on developed system	Lack of sufficient security in the developed product
Human factors	No clarity on functionality of different users	Complex user interface developed
		Nonuser friendly product developed
Specifications	Wrongly specified, underdeveloped system	Inappropriate product developed
		Obsolete product developed
		False productivity claims
Realistic	Unrealistic modification request from customer	Frequent change request due to bugs in the product developed
		Unrealistic change requirements
		Ambiguous improvement targets

engineering phases, risks can further be categorized into performance risks, cost risks, support risks, and schedule risks.

In general, there are many risks in the software engineering process and it is quite difficult to identify all of these. There are still some correlations between management risks versus financial risks, technical risks versus personal risks, business risks versus commercial risks, business risks versus economical risks and between financial risks versus commercial risks.

Since software engineering is a systematic development process, one must follow a proper sequence of steps irrespective of process model in use. None of the above-discussed classifications provides a clear and exhaustive categorization of risks in software projects.

It is, therefore, necessary to propose a new software risk classification where risks are classified based on the stage where they are identified. We propose a novel software risk classification approach where each category of risks follows a single stage of software development.

In this classification, risks identified in a single stage are grouped together. This classification will surely help the project managers to deal with the risks on that particular stage only where it was identified instead of delaying the mitigation till last stage. Following this classification, an early detection and early mitigation strategy can be adopted in order to minimize the loss in software project development.

Risk management strategies considered in most of the risk management models deal with handling generic risks related to the completion of the project and its analysis performed during requirement analysis only. These models highly focus on the requirement analysis and planning phase and try to identify the risks on generic sense. Since software project development is a systematic approach and it has clear distinct phases, a risk management strategy needs to be discussed at macro level where different risk management strategies may be adopted for individual phases of SDLC. Few of the above-discussed models also considered risks related to scheduling and quality during this phase. However, discussions on the risks related to design phase are yet to be considered by these risk models. There are few models [6, 11, 27] that focus on risks related to coding phase, whereas risk considerations related to testing and debugging are being discussed in [11, 25].

Along with generic risks, models as described by RISKIT, Hoodat_Rashidi, Danny also considered risks related to application phase and the risks related to maintenance phase are being discussed by SEI-SRE, RISKIT, PRORISK, ARMOR, SOFTRISK, PRM.

The existing risk management models barely consider software risks according to the phases of SDLC. SEI-SRE model focuses on the phases of SDLC where it emphasizes on the individual factors that may affect individual phases of SDLC.

Though SEI-SRE model is partially capable to deal with the practical aspects of IT projects, there is a limitation in modification of the template-based design. This may lead to inconclusive results for inexperienced user. It deals with risk from product, process, and constraints. It sets a risk baseline for a project. Unlike Boehm's work, SEI-SRE model did not clearly define the responsibilities of the project team.

Consequently, risk status data are not properly communicated between them. SEI-SRE has no method to analyze newly identified risks and it does not perform periodical review of risk treatment. This method only involves project personnel in its risk analysis and thus exclusion of stakeholders may create a possibility of incorrect requirements to the development team that incorporate some more risks which can lead to failure (Table 10).

Hence, a clear mapping is desired so as to understand which risk management model is best suited to follow the phases of SDLC. In our discussions a mapping is drawn in between the existing software risk management models and risks related to phases of SDLC. This mapping indicates a requirement of risk management model that deals risks at each phase individually rather than considering whole project at a time. An early detection of such risks is also essential so as to prepare the mitigation strategy well ahead of occurring of risks in reality.

Table 10 Risk models mapping with phases of SDLC

SN	Methods/models	Purpose	Risk element considered	Risks of which SDLC phases are considered
1	BOEHM [10]	Risk identification, analysis, prioritization, control	Generic risks and project-specific risks	Requirement analysis and planning
2	SoftRisk [20]	Risk identification, assessment, monitoring		Requirement and planning phase, maintenance phase
3	ARMOR [23]	Risk identification, analysis	All program module risks	
4	PRORISK [27], PRM [6]	Risk assessment, risk control	Software-related Generic risks	Requirement phase, coding phase, maintenance phase
5	RMM [18]	Risk assessment	Organizational risks	Not followed
6	ERM [42]	Risk identification, assessment	Generic risks and project-specific risks	
7	RAT [41]	Risk assessment, treatment and monitoring	Project risks of small and medium software	
8	TRM [13]	Risk analysis, mitigation	Team risks	
9	Agle et al. [14]	Risk handling	Risk related to team structure	
10	SEI-SRE [11]	Risk evaluation: Detection, specification, assessment, consolidation, mitigation	Product risks, process risks	Requirement phase, coding phase, testing phase, maintenance phase
11	SRAM [19]	Risk assessment, prioritization	Development risk	Requirement analysis
12	Armestrong [21]	Risk identification	Economic risk, business risk	
13	RISKIT [16]	Risk identification, analysis, monitoring, prioritize as per probability and impact	Generic risk, project risk, technical risk, schedule risk, business risk	Requirement phase, application and maintenance phase
14	H. Rashidi [25]	Risk measurement	Project risk, product risk, schedule risk, cost risk, quality risk, business risk	Planning phase, testing and debugging phase, application phase

(continued)

Table 10 (continued)

SN	Methods/models	Purpose	Risk element considered	Risks of which SDLC phases are considered
15	SERIM [29]	Risk assessment, risk ranking	Technical risk, cost risk, schedule risk, organizational risk, application risk	Requirement analysis and planning phase
16	RIMAM [28]	Risk identification, management, avoidance	Schedule risk and cost risk	
17	SRAEM [22]	Risk estimation	Technical risk, organization risk, environmental risk	
18	SRAEP [24]	Risk assessment, prioritization		
19	SERUM [17]	Implicit and explicit risk management	Generic risk, risk related to planning, development risk	
20	SPRMQ [7]	Risk factor identification, risk probability computation, effects on product quality, risk mitigation and monitoring	Product risks	
21	SITRAM [12]	Risk assessment	Technical risks	
22	Danny [26]	Risk mitigation	Operational risk	Application phase

5 Conclusions

The major contribution of this paper is the SDLC phasewise classification of risks that we have summarized in Sect. 3 using five tables and the consequent mapping of various risk models with different steps of SDLC described in Sect. 4. This systematic study followed by the proposed classification will open up a big horizon for entire risk management process. Researchers will now be able to apply various such models and analyze the occurrence and impact of risks at all steps of SDLC and can mitigate the risk once found meaningful. Generic tools might need to be developed for such purposes with option for tuning them for some specific business model.

References

1. Stern, R., Arias, J.C.: Review of risk management methods. Bus. Intell. J. **4**(1), 59–78 (2011)
2. Silva, P.S., Trigo, A., Varajão, J.: Collaborative risk management in software projects. In: Proceedings of the 8th International Conference on the Quality of Information and Communications Technology, pp. 157–160 (2012)
3. Guiling, L., Xiaojuan, Z.: Research on the risk management of IT project. In: Proceedings of International conference on E-Business and E -Government (ICEE), pp. 1–4, 6–8 May 2011

4. Tianyin, P.: Development of software project risk management model review. In: Proceedings of International Conference on AI, Management Science and Electronic Commerce, pp. 2979–2982 (2011)
5. Avdoshin, S.M., Pesotskaya, E.Y.: Software risk management. In: Proceedings of 7th Central and Eastern European Software Engineering Conference, Russia, pp. 1–6 (2011)
6. Westfall, L.: Software risk management. In: International Conference on Software Quality, San Diego, California, 8–10, February, 2011
7. Mofleh, H.M., Zahary, A.: A framework for software product risk management based on quality attributes and operational life cycle (SPRMQ). http://www.nauss.edu.sa/En/DigitalLibrary/Researches/Documents/2011/articles_2011_3102.pdf
8. Sarigiannidis, L., Chatzoglou, P.D.: Software development project risk management: a new conceptual framework. J. Softw. Eng. Appl. (JSEA) 4, 293–305 (2011)
9. Sathish Kumar, N., Vinay Sagar, A., Sudheer, Y.: Software risk management—an integrated approach. Global J. Comput. Sci. Technol. (GJCST) 10(15), 53–57 (2010)
10. Bohem, B.W.: Software risk management: principles and practices. IEEE Softw. 8, 32–41 (1991)
11. Carr, M.: Taxonomy-based risk identification. Software Engineering Institute, CMU/SEI-93-TR-6 (1993)
12. Loutchkina, I., Jain, L.C., Nguyen, T., Nesterov, S.: Systems' integration technical risks' assessment model (SITRAM). IEEE Trans. Syst. Man Cybern. Syst. 44(3), 342–352 (2014)
13. Higuera, R.P., Gluch, D.P., Murphy, R.L.: An introduction to team risk management. Special report CMU/SEI, SEI/CMU, Pittsburg, May 1994
14. Alge, B.J., Witheoff, C., Klein, H.J.: When does the medium matter? Knowledge building experiences and opportunities in decision making teams. Organ 91(1), 26–27 (2003)
15. Mead, N.R.: Measuring the software security requirements engineering process. In: Proceedings of 36th International Conference on Computer Software and Application Workshops, pp. 583–588 (2012)
16. Kontio, J., Basili, V.R.: Empirical evaluation of a risk management method. In: SEI Conference on Risk Management, Atlantic City (1997)
17. Greer, D.: SERUM: software engineering risk: understanding and management. J. Proj. Bus. Risk Manag. 1(4), 373–388 (1997)
18. Hillson, D.A.: Towards risk maturity model. Int. J. Proj. Bus. Risk Manag. 1(1), 35–45 (1997). ISSN:1366-2163 (Spring)
19. Foo, S.-W., Muruganatham, A., Software risk assessment model. ICMIT 2000, IEEE, pp. 536–544
20. Keshlaf, A.A., Hashim, K.: A model and prototype tool to manage software risks. In: Proceedings of the 1st Asia-Pacific Conference on Quality Software (APAQS'00), Washington, DC, USA (2000)
21. Armestrong, R., Adens, G.: Managing Software Project Risks. TASSC Technical paper, USA (2008). www.tassc-solutions.com. January 2008 Copyright 2001–2010, Tassc Limited
22. Gupta, D., Sadiq, M.: Software risk assessment and estimation model. In: International Conference on Computer Science and International Technology, pp. 963–967. IEEE Computer Society, Singapore (2008)
23. Rabbi, M., Mannan, K.: A review of software risk management for selection of best tools and techniques. In: Proceedings of 9th ACIS International Conference on Software Engineering, Artificial Intelligence, Networking, and Parallel/Distributed Computing, pp. 773–778 (2008)
24. Sadiq, M., Rahmani, M.K.I., Ahmad, M.W., Jung, S.: Software risk assessment and evaluation process (SRAEP) using model based approach. In: International Conference on Networking and Information Technology (ICNIT), pp. 171–177 (2010)
25. Hoodat, H., Rashidi, H.: Classification and analysis of risks in software engineering. In: WASET-2009, pp. 446–452
26. Danny, L.: Reducing operational risk by improving production software quality. Softw. Risk Reduction Rev. 13, 1–15 (2013)

27. Suebkuna, B., Ramingwong, S.: Towards a complete project oriented risk management model: a refinement of PRORISK. In: Eighth International Joint Conference on Computer Science and Software Engineering (JCSSE), pp. 349–354, 11–13 May 2011
28. Shahzad, B., Al-Ohali, Y., Abdullah, A.: Trivial model for mitigation of risks in software development life cycle. Int. J. Phys. Sci. 6(8), 2072–2082 (2011)
29. Roy, G.G.: A risk management framework for software engineering practice. In: Proceedings of the Australian Software Engineering Conference (AAWEC'04) (2014)
30. Amber, S., Shawoo, N., Begum, S.: Determination of risk during requirement engineering process. Int. J. Emerg. Trends Comput. Inform. Sci. 3(3), 358–364 (2012)
31. Pandey, D., Suman, U., Ramani, A.K.: Security requirement engineering issues in risk management. Int. J. Comput. Appl. 17(5), 11–14 (2011)
32. Islam, S., Houmb, S.H.: Integrating risk management activities into requirements engineering. RCIS-2010, Nice, France, May 2010, pp. 299–310
33. Drs. Erik, P.W.M.: Practical risk-based testing—product risk management: the PRISMA method, EuroSTAR-2011, Manchester, UK, pp. 1–24, 21–24 November 2011
34. Kwan, T.W., Leung, H.K.N.: A risk management methodology for project risk dependencies. IEEE Trans. Softw. Eng. 37(5), 635–648 (2011)
35. Lobato, L.L., Neto, S., da Mota, P.A., do Carmo Machado, I.: A study on risk management for software engineering. In: Proceedings of the EASE, pp. 47–51 (2012)
36. Lobato, L.L., da Mota, P.A., Neto, S., do Carmo Machado, I., de Almeida, E.S., de Lemos Meira, S.R.: Evidence from risk management in software product lines development: a cross-case analysis. In: Proceedings of 6th Brazilian Symposium on Software Components, Architectures and Reuse (2012)
37. Nolan, A.J., Abrahão, S., Clements, P.C., Pickard, A.: Requirements uncertainty in a software product line. In: Proceedings of 15th International Software Product Line Conference, pp. 223–231 (2011)
38. Lobato, L.L. et al.: Risk management in software product lines: an industrial case study. In: Proceedings of ICSSP, Switzerland, pp. 180–189 (2012)
39. Gonzalo, E., Gallardo, E.: Using configuration management and product line software paradigms to support the experimentation process in software engineering. RCIS-2012, Valencia, May 2012. pp. 1–6
40. Khoo, Y.B., Zhou, M., Kayis, B., Savci, S., Ahmed, A., Kusumo, R.: An agent-based risk management tool for concurrent engineering projects. Complex. Int. 12, 1–11 (2008)
41. Sharif, A.M., Rozan, M.Z.A.: Design and implementation of project time management risk assessment tool for SME projects using oracle application express. In: World Academy of Science, Engineering, and Technology (WASET), vol. 65, pp. 1221–1226 (2010)
42. Snekir, W.G., Walker, P.L.: Enterprise risk management: tools and techniques for effective implementation. Institute of management Accounts, pp. 1–31 (2007)

A New Service Discovery Approach for Community-Based Web

Adrija Bhattacharya, Smita Ghosh, Debarun Das
and Sankhayan Choudhury

Abstract In registry, services are identified by functionality but nonfunctional specifications (NFPs) should play an important role. Multiple registries are aggregated to offer a higher-level abstraction, called community. Community has its own functional description. In this paper, we have proposed an upgraded organization of registry as well as the community through the consideration of NFPs with the functional one. The newly proposed organization can be used to provide an efficient service discovery algorithm in term of execution time.

Keywords Web service · Service discovery · Nonfunctional parameters · Service community · Service registry

1 Introduction

Service providers publish web services within a registry along with description. Description of a service contains two parts; functional and nonfunctional. A service registry contains services with miscellaneous functional specifications. The services having analogous functionalities may be called "Similar services." But each of those similar services can have different nonfunctional specifications (npfs), as offered by the corresponding service providers. Due to the massive development of

A. Bhattacharya (✉) · S. Choudhury
University of Calcutta, Kolkata, India
e-mail: adrija.bhattacharya@gmail.com

S. Choudhury
e-mail: sankhayan@gmail.com

S. Ghosh · D. Das
Techno India Saltlake, Kolkata, India
e-mail: smita.ghosh.2710@gmail.com

D. Das
e-mail: idebarun@gmail.com

© Springer India 2016
R. Chaki et al. (eds.), *Advanced Computing and Systems for Security*,
Advances in Intelligent Systems and Computing 396,
DOI 10.1007/978-81-322-2653-6_10

139

web services, the registries are expanding day by day and as a result service discovery mechanism becomes complicated. The concept of community may be a solution from that aspect. In web technology, a community is a collection of registries that have common functionality [2]. A community can be viewed as domain specific functionality integration mediators representing the registry information [5]. The role of a community is to gather web services with similar functionalities (e.g., community hotel will contain services such as hotel reservation and hotel searching.). Thus this higher-level abstraction in form of a community helps in making the search of services in a more effective and organized way.

In general, in a registry the services are identified through the offered functionality. There is a work [1] that only considers the functionality of services during the community selection for a web registry. It first employs the idea of membership degree. In [4, 3], the main goal is to achieve the reliability of web services composition. This led to the requirement of a framework for this kind of composition that needed to satisfy handling errors and corresponding recovery actions. This requirement demanded rethinking over the idea of the community framework. In [6], discussion is made on an ontology-based infrastructure called METEOR-S. It helps in accessing a group of registries that are divided based on functionalities of various business domains and are clustered together into federations. Paper [2] presented an approach to help the service user for discovering web services. At first, the service discovery requirements are analyzed from the user's perspective. Then the description of a similarity measurement model for web services is proposed (by accessing data information from WSDL) and used for service discovery algorithm.

None of the existing works considered the nonfunctional property at the time of creating communities. Nonfunctional properties like location, cost, etc. are mentioned for a service but the existing discovery approaches are too focused on functional specifications. As a result, at the time of discovery, if the NFPs of the discovered service mismatches with the consumer's requirement, the discovery needs to be re-initiated and that in turn requires more execution time. Thus, we have proposed an updated organization of registry; a higher-level abstraction from the existing structure.

The service provider publishes the web services in the UDDI registries. Publishing the web services in the registry makes them available to the client. Providers have web registries at a specific standard called UDDI. A single provider can have multiple UDDI registries depending on its business operation. At the time of invoking, the service user can have to request directly to the provider. A user initiates the discovery mechanism by giving a query and receiving web services based on the preferences. Here, the consideration of nonfunctional parameters becomes essential. A community classically contains collection of information on the functional parameters of registries and registry contains a collection of service information [7]. However, in the proposed approach, each registry is evaluated for their memberships to community. Inclusion of nonfunctionality in the existing approach of membership degree is done.

We also consider the combinations of nonfunctional parameter while classifying the registry into communities. It is stored in such a way that the most occurring

NFPs with respect to communities and registries are reflected directly. This helps the discovery algorithm deliver the service efficiently in terms of accuracy and time.

The paper is organized in a following way. The proposed solution is discussed in detail in Sect. 2. Section 3 illustrates our proposal with an example. The corresponding discovery algorithm and its performance are mentioned in Sect. 4. The paper ends with the conclusion in Sect. 5.

2 Proposed Solution

In this section, we have described an updated organization of registry and community that in turn leads to an efficient service discovery. First, the structure is proposed for storing the services in a registry. Then an aggregation of registries into communities is done based on the available functional and nonfunctional information of services. Storing the service information in such a manner prevents traditional linear exhaustive search of services and returns the required services to a user quicker. The overall framework is depicted in Fig. 1 and the proposed solution is discussed in detail in the following subsections.

2.1 Registry Organization

In this work, we propose an alternative structure to store web services within registries. In the proposed structure, the web services are placed based on functional and NFPs. Here the registry is conceptualized as a multidimensional table where NFPs are considered as a dimension of the proposed table. For each nfp, the

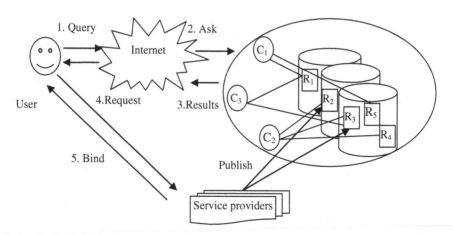

Fig. 1 Solution framework

nfp$_1$ Values / functionalities	v_{11}	...	v_{1kl}
fn1	Ser$_c$, Ser$_p$,...	...	------
fn2	--------	...	
fn3	Ser$_m$,...	...	------
....	....	....	...
Fnp	Ser$_j$	......	Ser$_u$,...

Fig. 2 A general view of proposed structure (with only one dimension)

services are stored as depicted in Fig. 2. Here each column indicates all possible value or range of values for a given nfp (for which the dimension is created), whereas a row implies a specific functionality. The intersection of a specific row and column indicates a service or a set of services that satisfy the typical combination of (nfp, fp) pair of values.

Let us consider a registry where the total numbers of nonfunctional and functional parameters are N and p, respectively and each nfp has a set of K values. Thus, the proposed registry organization should have N dimensions and each dimension will be conceptualized as table of $(k \times p)$ elements.

Thus as per the given proposal a registry is represented as a tuple of the form {F, nF} where F is broader Functionality (such as "Hotel," "Banking") and nF is set of all nonfunctional parameters. Here, nF = {nf$_i$} and nf$_i$ is the ith nonfunctional parameters of a particular service.

Calculation of membership degree

Each registry can be a part of a single community or more than one community. To determine which registry would be a part of which community, a membership degree (md) of each registry is calculated and is compared to a predetermined threshold value (th) [1]. The threshold value of community is fixed initially; gradually it can be increased or decreased. This md value helps in determining the belongingness of a given registry in a community. It determines the addition or deletion criteria of any registry from a community. The md value is calculated for each of the broader functionalities (e.g., as mentioned through set F in registry definition, for example, "Hotel") and importance is given to the NFPs also. Here md has two parts;

$$md = (md_f + md_{nf})/2,$$

where, md$_f$ is the membership degree with respect to only functionality and md$_{nf}$ is calculated considering the nfp's in the registry. md$_f$ is the same as calculated in [1] i.e., the ratio of occurrence of that functionality with respect to the total number of services in the registry.

$$md_f = \frac{\text{number of occurrences of functionality } (n)}{\text{total number of services in the registry } (T)}$$

Further in this approach, for a particular broader functionality, which nfp combination is the most frequent that is also calculated. This information is reflected within the membership calculation (in md_{nf}). For each of the registries, the nfp combinations (such as location and cost.) of a broader functionality are counted and ranked. The frequency of ith ranked combination is denoted by fr_i. The total rank sum is denoted by Rn.

Here,

$$Rn = \sum_{i=1}^{k} i + \sum_{j} tie_j * \text{rank},$$

where k is total number of combinations present in that registry, tie_j is jth tie and rank is the position where tie occurred. The weight (w_i) for ith nfp combination is calculated as,

$$w_i = \frac{Rn - (i - 1)}{Rn}.$$

Thus, md_{nf} is calculated for a particular broader functionality of a registry as,

$$md_{nf} = \sum_{i} (w_i * fr_i)/n.$$

Among all md values, the highest value and corresponding functionality is taken and checked if the community membership threshold is matched.

2.2 Defining the Community Schema

Communities are classified on the basis of a functionality paired with the occurring combination of related nonfunctional parameters in a given order. Every community is a collection of registries. Among these registries, one registry is chosen to be the master and the remaining registries become the slave registries. This classification is done in order to store the registries in the structure in a sorted manner of available services such that the searching of service through these registries can take place in descending order depending on the number of services in each unit in the community. The master is chosen in such a way that it has the maximum number of services available that has the functional description as well as most occurring nfp combination of a community. Rest of the services available for a community are present in the remaining slave registries. In a particular community, the registry with the highest md is the Master registry and the remaining registries are the slaves.

For a particular community with functionality F; Master Registry set (MR) = {F, nF_M}, ith Slave Registry set (SR$_i$) = {F, nF_{si}}, where i = 1, 2, 3, 4, ..., m. For each of all m slave registries possible combinations of parameters of nonfunctional parameter set nF_{si} are stored in another set called Com$_i$ = {all possible combinations of x|xε nF_{si}}. f_{nfp} is calculated by counting the frequency of every element in Com$_i$. The element in Com$_i$ for which f_{nfp} is maximum is named as freq_comb$_i$ for ith registry. All sets are combined together to form a set of prior nonfunctional combinations from slaves. This set is called Sl_nfp and defined as,

$$Sl\_nfp = \Lambda_{i=1 \text{ to } m}(freq\_comb_i).$$

The set of nonfunctional parameters for a particular community is finally computed as,

$$Co\_nfp = nF_M \text{ U } Sl\_nfp$$

The resultant set is then given by,

$$\{F, Co\_nfp\}.$$

It gives the appropriate description of the community which is described as the pairing of functionality and the most occurring combination of nonfunctional property. Hence, we find the most relevant nonfunctional parameters for a given functional parameter that represents a given community.

From the master registry the most occurring combination of nonfunctional parameters are determined. From the other slave registries, a set is constructed containing the intersection of nfp combinations present in all of them. This set of nfp combination, generated from the slave registries, is joined (union) with the nfp combination set of the master and the resultant set determines the classified description of a particular community.

It becomes difficult to determine the final number of combinations of NFPs to be specified in the community schema definition. After classification of community, each community is described as a service container where all the services have same functionality along with the most occurring combination of nonfunctionalities. The main community schema is the combination of all communities with the most relevant combination of nonfunctionalities corresponding to each of the functionalities. The conceptual schema structure is depicted in Fig. 3. This relevant combination is gathered from each community description i.e., the super set of the combinations is gathered to form each tuple of the community structure. Now the set of all valid NFPs in community schema (Sc_nfp) is determined by

Community name	Functionality	Nfp combination$_1$	.....	Nfp Combination$_t$
C_1	Hotel	R_3, R_7		
C_2	Banking	R_9		R_4, R_8
....		----		
C_M	Healthcare	-----		R_1

Fig. 3 Community schema

$$Sc\_nfp = \bigcup_{i=1}^{M} Co\_nfp_i.$$

Total number of columns for NFPs in the community schema is same as the number of elements in Sc_nfp.

2.3 Organization of a Community

The community structure is redesigned according to the modified registry organization. This subsection describes the method of constructing the proposed community. The most important issue for building a community is to take the decision regarding aggregation i.e., the selection of registries to be grouped within a community. The next step is to define a suitable community description including nonfunctional parameters. This helps in classifying and accessing community information by not only its similar functionality but also a combination of fp and NFPs.

The proposed community organization is a three-dimensional structure with community id (based on specific functionality), the list of combination of nonfunctional parameters, and the array of registries as dimension. The first dimension indicates the community id. Community id is derived as a particular higher abstraction of common functionalities that registries hold. Let us consider that the registries are populated with various services having functionalities such as HotelBooking, HotelBrowsing, etc. Then a community may be identified as "hotel" containing registries that have some significant number of hotel services. The second dimension is the list of the occurring combination of nonfunctional properties for the services in that community in a specific order. For the hotel community, availability, cost, and locations are treated as some of the nonfunctional parameters those are valid for hotel services. A set of occurring parameters from the existing NFPs are identified from the registry entries and are included within community organization. The third dimension is represented by a linear structure that stores the information of registries specifying the corresponding nonfunctional parameter combination in a sorted order (based on maximum satisfying requests to fewer). These newly included information within community organization will be helpful for getting a quick match for popular (fp, nfp) pairs in context of a domain referred as a community. The above said description is represented in Fig. 4.

Fig. 4 Proposed structure of
a web service community

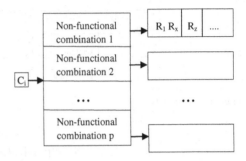

2.4 Membership Alteration and Threshold Adjustment

A community is associated with a threshold value. The membership degree value of
each functionality in the registry is compared with a threshold value (th) of the
community. If md is more than th then the registry gets qualified to be a part of that
community. One registry may be part of more than one community as long as its
md is greater than the threshold values of those communities. Any change in the md
value also leads to alteration of the membership of a particular registry in a community. Thus, a regular monitoring is required to keep the communities updated.

Whenever a new service is added to a registry or deleted from it, it leads to a
change in the total number of services in a given registry; the membership degree
(md) which determines the community to which a particular registry belongs to,
also change. As mentioned before, a registry may be a part of more than one
community. Thus, addition or deletion of new service may result in adding the
particular registry to a community, deleting the same from another community, both
or none. Thus, a regular monitoring is required to keep the membership parameter
value (md) updated in order to ensure efficient grouping of registries into respective
communities. As the membership degree (md) of a registry decides the community
to which a particular registry will belong, any changes in the md value may alter the
membership of a particular registry in a community and also affects the ranking of
that particular registry in the community. Addition of services or deletion of the
same or both may result in the change of the membership degree of the registry. The
corresponding algorithm is given as follows.

Membership alteration Algorithm
Any change is the md value of a registry may result in the following:

- addition or deletion of the registry from a community
- change in master–slave relation

To manage the stability through the process of services registry modification and
membership alteration issues identified above, an algorithm is defined here,

> *Check if $(md_{new}(R_i) > th_{ci})$*
> *then R_i belongs to community C_i*
> *else if $(md_{new}(R_i) < th_{ci})$*
> *then R_i is removed from community C_i*
> *if (R_i is slave registry)*
> *then continue*
> *else if (R_i is master registry)*
> *then calculate the next highest md in community*
> *declare new master registry.*

where $md_{new}(R_i)$ is the new membership degree value of register R_i, after addition or deletion of services in registry R_i. th_{ci} is the predefined threshold value of community C_i; it is a fraction.

Initially the threshold for each community is fixed at 0.5. At the first revision, each of the community is checked from community schema. If there is more than one entry in most of the cells in a row; then that community threshold is increased such that less than half of the cell in a row of community schema contains more than one value at a time. Similarly, if there are too small number of registries in a row of community schema; that corresponding community threshold is decreased, such that at least half of the cells are filled by one or more registry names. This procedure can be repeated depending on the frequency of updating the registries.

3 Illustration with an Example

In this section, we have illustrated our proposal regarding registry and community and have tried to show the utility of the upgraded community definition for service discovery. The registry structure is redefined based on the available service information. Let us consider a registry named R_1. Conventionally the registry contains the information about the services in the form of functional description, nonfunctional description, access information, etc. Each row of R_1 contains services with varying functionalities; such as HotelBooking, HotelSearch, CheckAvailability, CreditcardPayment, AccountOpen, and BalanceEnquirey. There exist multiple services with respect to single functionality. The nonfunctional parameters of those services can vary widely. Here, we assume the nonfunctionalities with respect to hotel-related services (HotelBooking, HotelSearch, and CheckAvailability) are cost, reliability, and location. Similarly the nonfunctional parameters of bank-related services (CreditcardPayment, AccountOpen, and BalanceEnquirey) can be reliability, security, and providers. Thus, considering all service functionalities and corresponding nonfunctionalities under registry R_1 two community (broader functionality, with three functionalities each) and five nonfunctional parameters are identified. Two functionalities of communities are "hotel" and "banking" and cost, location, reliability, security, and provider are nonfunctional parameters. Now according to our proposed approach the registry R_1 is reconstructed with distributing the services according to five nonfunctional parameters in five different tables (as illustrated in Fig. 2). Each of the table contains six rows with respect to six functionalities. These five tables as a whole represent a single registry R_1. R_1 can be member of two communities at a time; provided the membership degree of two broader functionalities hotel and banking exceeds the threshold value of the communities "Hotel" and "Banking," respectively.

Another rearrangement in community organization is proposed. The task is to include the occurring combinations of nonfunctional parameters in a sorted order with respect to a specific functionality in community organization. For example, the hotel-related services may have the combination of cost and reliability as most occurring. So, in community "hotel" the registry R_1 occurs with respect to combination of cost and reliability (such as in Fig. 4). Similarly most occurring nonfunctional parameters for "banking" may be the providers. So, R_1 occurs in community "banking" with NFP "provider." Many registries except R_1 may be placed within community "hotel" according to the occurring nfp combinations within those registries. Among these, registries say a registry R_{master} has highest md value and as a result this is considered as the master registry of "hotel." Now say this community name is C_p. C_p has a master R_{master} and other slave registries. The occurring set of nfp combinations are identified for C_p. The most occurring nfp combination set is constructed by rules as discussed in Sect. 2.2. C_p is defined within the community schema with a set of nonfunctional combinations and corresponding registry names (as in Fig. 3).

When a particular query in the form Q(f, nf) occurs, the functional portion f and nonfunctional part nf (combination of nonfunctional parameters) is matched with community schema entries first. From that schema, the corresponding master and slave registries are found readily. Then the search procedure starts from master registry and continues as described in Sect. 4. Suppose a query is in the form Q ("HotelBooking," cost = "<Rs. 2000," and location = "Delhi"). Then at first, the community schema is searched with brokader functionality "hotel" and nfp combination (cost and location). It will find out some registries like R_5, R_8, etc. It will then search for the functionality "HotelBooking" within R_5 with mentioned cost and location; if found, then it will return the result. Otherwise it will search in the subsequent slave registries (R_8 and so on).

4 Service Discovery Algorithm and Analysis

Initially, the user gives a query of the form Q(f, nf), where f represents the functionality of the service requested and nf is the set of nonfunctional parameter values that are given as preferences by the user. Based on the functionality f, the corresponding community is selected and based on nf, a particular group of registries is selected within that community. In the registries that are selected, the required services are returned to the user, based on the services that satisfy both the functionality and the combination of nonfunctional parameter required by the user. The service discovery algorithm runs in two parts; first is overall discovery mechanism and the other part is discovery within registries. The algorithms are described in detail as follows

Algorithm for Overall Service Discovery

Input: User Query Q(fn, {nf}), Community Schema Structure, Core Community Structure CCSt, Registries R1, R2, R3,...,Rn

   1. Ck → Community matched on CSSt based on fn

   2 If {nf} ε CSSt

   3. do

   4.   array Services → searchRegister(Ri, Q(fn,{nf})) from CSSt

   refer algorithm for dioscovery within a registry

   5. done

   6. else

   7. do

   8.   array Services → searchRegister(Ri, Q(fn,{nf})) from CCSt of Community Ck
......*refer Algorithm 5.2*

   9. done

   10. return array Services

The search of web services in a particular registry does not follow the normal linear search. The algorithm for service search within a particular registry is given as follows.

Algorithm for Service discovery within a registry

   1.functionsearchRegistry(R_k,Q(f,nf))

   2.do

   3.   np →Number of non-functional parameters in nf

   4. k_h →Number of values of nfp_h

   5.   tn →Total number of nfps in the registry

   6.   for i →1to np

   7.   do

   8.   For j →1to$\prod_{h=1 to np}(k_h)$

   9.   do

   10.   ind →required index in Registry R_k

   11.   array Services → Set of matched services at index ind in R_k

   12.   done

   13.   done.

   14.   Return array Services

   15. done.

Table 1 Comparison between liner and nonlinear discovery algorithm

No. of services	Running time using proposed method (ms)	Running time using baseline search (ms)
500	0.789	0.602
1000	1.305	1.570
2000	1.434	2.710
4000	1.587	4.797
8000	1.598	9.698
16000	1.779	42.600

The time complexity for the search of services is as follows:

$$\text{Time Complexity} = O(\text{comb\_nfp} * \text{cols})$$

$$= O\left({}^{tn}C_{np} * \prod_{h=1 \text{ to } np} (k_h)\right)$$

where, comb_nfp $\rightarrow$ Total number of combinations when we take np number of registries out of a total of tn nonfunctional parameters in the registry.

cols $\rightarrow$ Total number of columns for the given combination in the proposed structure. Thus, we can comment that due to the proposed structure of registry, the running time does not increase significantly with the increase in the number of services in contrast to linear search.

Here, we compare our method with a baseline approach that uses elaborate search to check all services with required functionality (as mentioned in query) and then matching nonfunctional parameters for best possible match. We measure the efficiency of two methods by the search time.

As shown in the Table 1, we see that although the running time obtained using our method for small number of services (500) is greater than the running time obtained using linear search, yet we see that with the increase in the number of services, our running time increases negligibly when compared to the increase in running time using linear search. We represent the running time of the two ways of web service discovery in a bar graph as shown in Fig. 5.

Fig. 5 Comparison between the running time of proposed method and by baseline mechanism

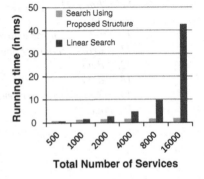

5 Conclusion

In this proposed work, our overall motivation is to include the nonfunctional parameters within community schema that in turn offers a better service discovery approach. Moreover, the inclusion principle of a registry within community is modified, whereas the functionalities for the available combinations of NFPs are considered as a whole for inclusion criterion. As a result the master registry of a community will be able to specify the services those are needed to satisfy user query in most of the cases. Here the registries are arranged in a sorted order within community. The community schema includes the information of registries with the most occurring nonfunctional parameter combinations. This actually reflects the most common choice of NFPs by providers. If the query demands any of the most common NFPs; then the name of the registry(s) containing that specification is readily available from the community schema. In case of a query that demands less common set of NFPs, the discovery method finds the registry name of corresponding nfp combination from the community structure instead of searching linearly within community as baseline method. The claim is being established through the experimental result also.

References

1. Bouchaala, O., Sellami, M., Gaaloul, W., Tata, S., Jmaiel, M.: Modeling and managing communities of web service registries. In: WEBIST, pp. 88–102 (2011)
2. Liu, X., Huang, G., Mei, H.: Discovering homogeneous web service community in the user-centric web environment. IEEE Trans. Serv. Comput. 2(2), 167–181 (2009)
3. Maamar, Z., Benslimane, D., Thiran, P., Ghedira, C., Dustdar, S., Sattanathan, S.: Towards a context-based multi-type policy approach for Web services composition. Data Knowl. Eng. **62**(2), 327–351 (2007)
4. Maamar, Z., Subramanian, S., Thiran, P., Benslimane, D., Bentahar, J.: An approach to engineer communities of web services: concepts, architecture, operation, and deployment. Int. J. E-Bus. Res. (IJEBR) **5**(4), 1–21 (2009)
5. Paik, H.Y., Benatallah, B., Toumani, F.: Toward self-organizing service communities. IEEE Trans. Syst. Man Cybern. Part A Syst. Hum. **35**(3), 408–419 (2005)
6. Sivashanmugam, K., Verma, K., Sheth, A.: Discovery of web services in a federated registry environment. In: Proceedings of the IEEE International Conference on Web Services, pp. 270–278. IEEE (2004)
7. Vijayakumar, D., Mahmoud, Q.H.: A framework for the service provisioning of community-contributed web APIs. In: 27th Canadian Conference on Electrical and Computer Engineering (CCECE), pp. 1–7. IEEE (2014)

Data-Flow Analysis-Based Approach of Database Watermarking

Sapana Rani, Preeti Kachhap and Raju Halder

Abstract In this paper, we propose a persistent watermarking technique of information systems supported by relational databases at the back-end. The persistency is achieved by identifying an invariant part of the database which remains unchanged w.r.t. the operations in the associated applications. To achieve this, we apply static data-flow analysis technique to the applications. The watermark is then embedded into the invariant part of the database, leading to a persistent watermark. We also watermark the associated applications in the information system by using opaque predicates which are obtained from the variant part of the database.

Keywords Persistent watermarking · Relational databases · Data-flow analysis · Security

1 Introduction

Database watermarking of relational databases has received much attentions to the research community over the last decade when various application scenarios, e.g., database-as-a-service, data-mining technologies, online B2B interactions, etc., demand an effective way to protect database information from various fraudulent activities, like illegal redistribution, ownership claims, forgery, theft, etc. [15, 26]. Figure 1 depicts a pictorial view of database watermarking techniques, where a watermark W is embedded into the original database using a private key K (known only to the owner) and later the verification process is performed on any suspicious

S. Rani (✉) · P. Kachhap · R. Halder
Indian Institute of Technology, Patna, India
e-mail: sapana.pcs13@iitp.ac.in

P. Kachhap
e-mail: preeti.cs10@iitp.ac.in

R. Halder
e-mail: halder@iitp.ac.in

© Springer India 2016
R. Chaki et al. (eds.), *Advanced Computing and Systems for Security*,
Advances in Intelligent Systems and Computing 396,
DOI 10.1007/978-81-322-2653-6_11

153

database using the same private key K by extracting and comparing the embedded watermark (if present) with the original watermark information.

1.1 Related Works

Existing watermarking techniques are categorized into two: distortion-based and distortion-free. Distortion-based techniques [1, 10, 11, 25, 27, 28] introduce distortion to the underlying database data, and hence, usability is a prime concern while watermarking. Distortion should always be introduced in such a way that it is tolerable and does not destroy the usability of the data at all. Watermarking in [1] is performed by flipping bits in numerical values at some predetermined positions based on the secret parameters. Image as watermark is embedded at bit-level in [28]. Approaches in [10, 27] are based on database content: The characteristics of database data is extracted and embedded as watermark into itself. Authors in [11] proposed a reversible-watermarking technique which allows to recover the original data from the distorted watermarked data. Khanduja et al. [19] proposed a secure embedding of blind and multi-bit watermarks using Bacterial Foraging Algorithm. Later, they used voice as biometric identifier for watermarking [18]. Unlike numerical values, categorical data type and nonnumeric multi-word attributes are also considered as cover for watermarking in [2, 25]. Distortion-free watermarking techniques [5, 6, 13, 20, 21], on the other hand, do not introduce any distortion. Unlike distortion-based techniques, watermark is generated from the database rather than embedding. In [4, 21], hash value of the database is extracted as watermark information. Approaches in [5, 6, 20] are based on the conversion of database relation into a binary form to be used as watermark. In [17], watermark is generated based on digit frequency, length of data values, etc. in the database, whereas [7] generates the watermark based on the grouping of data into square matrix and the computation of determinant and diagonals' minor for each group. Although the approach [7] is not economically viable, but suitable to detect multifaceted attacks and is resilient against tuples insertion-deletion attack and value modification attack.

1.2 Motivations

This is to be observed that most of the distortion-based techniques in the literature use a part of the database content as cover [10, 27, 28], and therefore, a number of update or delete operations may distort the watermark or may make the watermark undetectable. Also re-watermarking the database is very expensive process. Authors in [12, 13] first address a key issue, called *persistency*, in the context of database watermarking where database tuples are being updated or deleted frequently by the associated legitimate applications. Their approaches aim at preserving persistency

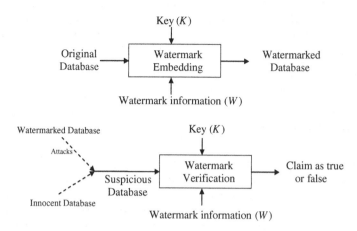

Fig. 1 Basic watermarking technique

of the embedded watermark under usual database operations: watermark is embedded in an invariant part of the database (w.r.t. database operations), while the same is generated from the abstract variant part representing properties instead of actual values. However, they did not specify any approach to identify the variant/invariant part while watermarking a complete information system consisting of a set of applications interacting with a database at the back-end.

1.3   Contributions

In this paper, we propose a data-flow analysis-based approach which serves as a generic framework for persistent database watermarking. Unlike existing approaches, we consider watermarking of a complete information system which includes both the back-end database and the associated applications legitimately accessing or manipulating the data in the database. In particular, our proposal is unfolded into the following phases:

- Formulation of data-flow equations for the applications embedding query languages.
- Analysis of the applications based on the data-flow equations which effectively identifies an invariant part of the underlying database instances.
- Watermarking of the invariant part by distortion-based technique.
- Generation of Opaque Predicates from the variant part respecting the integrity constraints of the database systems.
- Embedding opaque predicates as watermarks into the associated applications.

The structure of the paper is as follows: Sect. 2 provides a motivating example. Section 3 recalls some basic notions about persistent watermarking, data-flow analysis, etc. The proposed technique is discussed in Sect. 4. In Sects. 5 and 6, we provide, respectively a brief discussion on the complexity and robustness of our proposal. Experimental results are presented in Sect. 7. Finally, we draw our conclusions in Sect. 8.

2 Running Example

Consider, three online trading companies, say x, y, z, who are maintaining their own databases and the associated applications. Figure 1 depicts one such database which stores the details of the customers, various products, and the purchase history. Suppose, three companies have decided to collaborate, aiming at making the online purchasing system more attractive to the customers in terms of product availability.

However, according to the policy, each company can perform, in addition, its own business independently. A common interface after collaboration is developed and is allowed to access any of the three databases. This makes the database information vulnerable to various kinds of attacks, e.g., theft, illegal redistribution, ownership claiming, etc. Therefore, it is mandatory to watermark individual database in order to prevent above mentioned attacks.

Consider a code-fragment[1] P depicted in Fig. 2 which accesses and manipulates database of Table 1. The code either inserts order details (statement 7–11) or offers gifts to the premium customers (statement 13–16). This is to be noted that the database part corresponding to the attributes "TotalAmt" and "Offer" can possibly be updated by the application—hence it is a variant part. The rest of the database acts as invariant part. This is immediate that any watermark embedded into this variant part may get destroyed or undetectable due to the legitimate update operations on the values.

In the subsequent sections, we propose an efficient way to identify invariant and variant part of the underlying databases w.r.t. the associated applications in the system. This will enhance the existing watermarking techniques w.r.t. the persistency issue.

3 Basic Concepts

In this Section, we recall some basic notion about persistent watermarking from [13].

Persistent watermark Given a database dB and a set of associate applications A, we denote by $\langle dB, A \rangle$ an information system model. Let d_0 be the initial state in which the watermark W is embedded. When applications from A are processed on

---

[1]Observe that we do not follow any specific language syntax.

```
0.   start;

1.   Statement stmt=DriverManager.getConnection("jdbc:mysql://localhost:3306/demo","root","tiger").
     createStatement();

2.   $choice = read();

3.   $Item = read();

4.   $Item_count = read();

5.   $Cust_id = read();

6.   if($choice == "purchase"){

7.   $rs1 = SELECT ItemNo, UnitPrice FROM Store WHERE item=$Item and NoAvail>0;

8.   if($rs1.next()){

9.   $Ord_no = generate();

10.  INSERT INTO Order(OrderId, CustomerId, ItemNum, count, date, offer) VALUES ($Ord_no,
     $cust_id, $rs1.ItemNo, $Item_count, today(), NULL);

11.  UPDATE Customer SET TotalAmt = TotalAmt + $Item_count * $rs1.UnitPrice WHERE CustId
     = $Cust_id;}}

12.  if($choice == "offer"){

13.  $rs2 = SELECT CustId FROM Cust WHERE TotalAmt>5000;

14.  while($rs2.next()){

15.  $gift = read();

16.  UPDATE Order SET offer = $gift WHERE CustomerId = $rs2.CustId;}}

17.  stop;
```

Fig. 2 Program *P*

Table 1 Online trading database

(a) Table "cust"

CustId	CustName	Address	Age	TotalAmt
1001CI01	Rachel	London	22	2000
1001CI02	Albert	New York	25	7000
1001CI03	John	Japan	27	4500

(b) Table "Store"

ItemNo	ItemName	NoAvail	UnitPrice
TN01	Notebook	23	200
TN02	Calculator	25	1000

(c) Table "Order"

OrderId	CustomerId	ItemNum	Count	Date	Offer
111OI01	1001CI02	TN02	2	2-12-2012	NIL
111OI02	1001CI01	TN02	1	4-1-2013	NIL

d_0, the state changes and goes through a number of valid states $d_1, d_2 ..., d_{n-1}$. The watermark W is persistent if we can extract and verify it blindly from any of the following $n - 1$ states successfully.

Definition 1 (*Persistent Watermark*)

Let $\langle \text{dB}, A \rangle$ be an information system model where A represents the set of associated applications interacting with the database dB. Suppose the initial state of dB is d_0. The processing of applications from A over d_0 yields to a set of valid states $d_1, ..., d_{n-1}$. A watermark W embedded in state d_0 of dB is called persistent if

$$\forall i \in [1..(n - 1)], \text{ verify}(d_0, W) = \text{verify}(d_i, W)$$

where verify(d, W) is a boolean function such that the probability of "verify(d, W) = true" is negligible when W is not the watermark embedded in d.

Variant versus Invariant Database Part Consider an information system $\langle \text{dB}, A1 \rangle$ where A is the set of applications interacting with database dB. For any state d_i, $i \in [0...(n - 1)]$, we can partition the data cells in d_i into two parts: Invariant and Variant. Invariant part contains those data cells that are not updated or deleted by the applications in A, whereas data cells in variant part of d_i may change under the processing of applications in A.

Let CELL_{d_i} be the set of cells in the state d_i. The set of invariant cells of d_i w.r.t. A is denoted by $\text{Inv}_{d_i}^A \subseteq \text{CELL}_{d_i}$. For each tuple $t \in d_i$, the invariant part of t is $\text{Inv}_t^A \subseteq \text{Inv}_{d_i}^A$. Thus, $\text{Inv}_{d_i}^A = \bigcup_{t_j \in d_i} \text{Inv}_t^A$. The variant part w.r.t. A, on the other hand, is defined as $\text{Var}_{d_i}^A = \text{CELL}_{d_i} - \text{Inv}_{d_i}^A$.

Data-flow Analysis Data-flow analysis is a technique for gathering information about the dynamic behavior of programs by only examining the static code [24]. A program's control-flow graph (CFG) is used to define data-flow equations for each of the nodes in the graph. Data-flow analysis can be performed either in a forward direction or in a backward direction, depending on the equations defined. The least fix-point solution of the equations provides the required information about the program. The information gathered is often used by compilers when optimizing a program. A canonical example of a data-flow analysis is reaching definitions.

4 Proposed Technique

The intuition of our proposal is to make the embedded watermark persistent w.r.t. all possible operations in the information system. As database states change frequently under various legitimate operations in the associated applications, the content dependent watermarks embedded into the database are highly susceptible to benign updates. In particular, update and delete operations may remove or distort any existing watermark of the database [10, 27, 28].

In order to make the watermark persistent, our proposal aims at identifying some invariant parts of the database states which remain unchanged w.r.t. the applications. To this aim, we apply static data-flow analysis technique to the associated applications which identifies various parts of the database, called variant parts, targeted by update, or delete operations in the applications. The complement of this variant part in the database acts as invariant part and is used for persistent watermarking. For instance, any database part retrieved by SQL select statement remains unchanged and is, of course, suitable for persistent watermarking. We also watermark the associated applications in the information system by using opaque predicates obtained from the variant part.

Summarizing, the proposed technique consists of the following phases:

- Identifying variant and invariant parts of the database, by performing data-flow analysis to the associated applications.
- Watermarking of invariant database parts.
- Watermarking of associated applications by using opaque predicates obtained from the variant part.

4.1 Data-Flow Analysis

In this phase, we analyze the associated applications based on the data-flow equations in order to collect information about the part of the database information updated or deleted at each point of the applications.

The data-flow equations for various commands in the applications embedding query languages are defined in Fig. 3. The abstract syntax of update and delete statements are denoted by $\langle \vec{v}_d \overset{\text{upd}}{=} \vec{e}, \ \phi \rangle$ and $\langle \text{del}(\vec{v}_d), \ \phi \rangle$ respectively, where $\vec{v}_d = \langle a_1, a_2, \ldots, a_r \rangle$ denotes a sequence of database attributes, $\vec{e} = \langle e_1, e_2, \ldots, e_r \rangle$ denotes a sequence of arithmetic expressions, and ϕ denotes the WHERE-part of the statements following first-order formula [14]. We denote by notations $\text{upd}(\vec{v}_d)|_\phi$ and $\text{del}(\vec{v}_d)|_\phi$ the part of the database updated and deleted by $\langle \vec{v}_d \overset{\text{upd}}{=} \vec{e}, \ \phi \rangle$ and $\langle \text{del}(\vec{v}_d), \ \phi \rangle$ respectively. Observe that any database part is identified by a subset of attributes $\vec{v}_d$ values corresponding to a subset of tuples satisfied by ϕ. The notation (x, n) represents that x is defined at program point n, whereas $(x, ?)$ represents that x is defined by any program point. In case of conditional node with boolean expression b, we denote by notation $\text{JOIN}(n)|_b$ the information restricted by b.

The data-flow analysis is performed by using data-flow equations for each node of the control-flow graph and solves them by repeatedly calculating the output from the input locally at each node until the whole system stabilizes, i.e., it reaches a fix point. The least fix-point solution of the equations provides the information about the variant part of the database possibly updated or deleted by the program. Observe that during solving the data-flow equations, the result in any iteration may contain

Fig. 3 Data-flow equations
of applications embedding
query languages

Assignment node n.
$[\![n]\!] = (\text{JOIN}(n) \setminus \{(x,?)\}) \cup \{(x,n)\}$

Conditional node n.
$[\![n]\!] = \text{JOIN}(n)|_b$

UPDATE node n.
$[\![n]\!] = \text{JOIN}(n) \cup \{(\vec{v_d}|_\phi, n)\}$
$\qquad = \text{JOIN}(n) \cup \{(a_1|_\phi, n), (a_2|_\phi, n), \ldots, (a_r|_\phi, n)\}$

DELETE node n.
$[\![n]\!] = \text{JOIN}(n) \cup \{(\vec{v_d}|_\phi, n)\}$
$\qquad = \text{JOIN}(n) \cup \{(a_1|_\phi, n), (a_2|_\phi, n), \ldots, (a_r|_\phi, n)\}$

Other nodes.
$[\![n]\!] = \text{JOIN}(n)$

where $\text{JOIN}(n) = \bigcup_{w \in \text{pred}(n)} [\![w]\!]$.

multiple definitions of the same attributes corresponding to different conditions (for example, say $\vec{v_d}|_{\phi_1}$ and $\vec{v_d}|_{\phi_2}$).[2] In such case, we use merge function defined below:

$$\text{merge}((a|_{\phi_1}, n_1), (a|_{\phi_2}, n_2)) = (a|_{\phi_1 \vee \phi_2}, \{n_1, n_2\})$$

This yields a modified data-flow equations for UPDATE and DELETE as follows:

UPDATE node n.
$[\![n]\!] = \text{merge}\left(\text{JOIN}(n) \cup \{(\vec{v_d}|_\phi, n)\}\right)$
$\qquad = \text{merge}\left(\text{JOIN}(n) \cup \{(a_1|_\phi, n), (a_2|_\phi, n), \ldots, (a_r|_\phi, n)\}\right)$

DELETE node n.
$[\![n]\!] = \text{merge}\left(\text{JOIN}(n) \cup \{(\vec{v_d})|_\phi, n)\}\right)$
$\qquad = \text{merge}\left(\text{JOIN}(n) \cup \{(a_1|_\phi, n), (a_2|_\phi, n), \ldots, (a_r|_\phi, n)\}\right)$

Lattice Structure Defining Data-flow. Let Lab, Var, ψ be the set of program points, the set of program variables and the set of well-formed formulas (in first-order logic), respectively. Let $R = \text{Var} \times \psi \times \wp(\text{Lab})$. The Lattice is defined as $(\wp(R), \subseteq, \varnothing, R, \cup, \cap)$, where $\varnothing$ is the bottom element and R is the top element of the lattice. The lowest upper bound $\cup$ is defined as:

---

[2]By notation $\vec{v_d}|_\phi$ we denote the part of the database corresponding to the attributes $\vec{v_d}$ and tuples satisfying the condition ϕ.

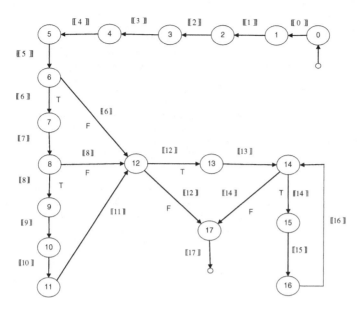

Fig. 4 Control-flow graph of P

$$\{(x_i, \phi_i, \{l_{i,m}\})\} \cup \{(x_j, \phi_j, \{l_{j,n}\})\} = \begin{cases} \{(x_i, \phi_i \vee \phi_j, \{l_{i,m}\} \cup \{l_{j,n}\})\} \\ \{(x_i, \phi_i, \{l_{i,m}\})(x_j, \phi_j, \{l_{j,n}\})\} \end{cases}$$

and the greatest lower bound $\cap$ is defined as:

$$\{(x_i, \phi_i, \{l_{i,m}\})\} \cap \{(x_j, \phi_j, \{l_{j,n}\})\} \begin{cases} \{(x_i, \phi_i \wedge \phi_j, \{l_{i,m}\} \cap \{l_{j,n}\})\} \text{ if } x_i = x_j \\ \emptyset \qquad \text{otherwise} \end{cases}$$

Example 1 Let us illustrate the data-flow analysis on the running example P of Sect. 2. The control-flow graph of P and the data-flow equations for each node are depicted in Figs. 4 and 5[3] respectively. If we solve the equations assuming the initial value as empty set, we get the least fix-point solution depicted in Fig. 6. The solution clearly indicates that the data corresponding to the attributes "TotalAmt" and "Offer" may possibly be defined at program points 11 and 16. Therefore, this part act as variant part of the database, while the remaining acts as an invariant part.

---

[3]For the sake of simplicity, we omit set-curly-braces incase of singleton set.

Fig. 5 Data-flow equations
of control-flow graph nodes
of *P*

$[\![0]\!] = \{\}$

$[\![1]\!] = ([\![0]\!]\backslash\{(\text{stmt}, ?)\})\cup\{(\text{stmt}, 1)\}$

$[\![2]\!] = ([\![1]\!]\backslash\{(\$\text{choice}, ?)\})\cup\{(\$\text{choice}, 2)\}$

$[\![3]\!] = ([\![2]\!]\backslash\{(\$\text{Item}, ?)\})\cup\{(\$\text{Item}, 3)\}$

$[\![4]\!] = ([\![3]\!]\backslash\{(\$\text{Item\_count}, ?)\})\cup\{(\$\text{Item\_count}, 4)\}$

$[\![5]\!] = ([\![4]\!]\backslash\{(\$\text{Cust\_id}, ?)\})\cup\{(\$\text{Cust\_id}, 5)\}$

$[\![6]\!] = [\![5]\!]|_{\$\text{choice}==\text{"purchase"}}$

$[\![7]\!] = ([\![6]\!]\backslash\{(\$\text{rs1}, ?)\})\cup\{(\$\text{rs1}, 7)\}$

$[\![8]\!] = [\![7]\!]|_{\text{rs1.next()}}$

$[\![9]\!] = ([\![8]\!]\backslash\{(\$\text{Ord\_no}, ?)\})\cup\{(\$\text{Ord\_no}, 9)\}$

$[\![10]\!] = [\![9]\!]$

$[\![11]\!] = [\![10]\!]\cup\{\text{upd(TotalAmt)}|_{\text{WHERE CustId = \$Cust\_id}}\}$

$[\![12]\!] = ([\![6]\!] \cup [\![8]\!] \cup [\![11]\!])|_{\$\text{choice}==\text{offer}}$

$[\![13]\!] = ([\![12]\!]\backslash\{(\$\text{rs2}, ?)\})\cup\{(\$\text{rs2}, 13)\}$

$[\![14]\!] = ([\![13]\!] \cup [\![16]\!])|_{\text{rs2.next()}}$

$[\![15]\!] = ([\![14]\!]\backslash\{(\$\text{gift}, ?)\})\cup\{(\$\text{gift}, 15)\}$

$[\![16]\!] = [\![15]\!]\cup\{\text{upd(offer)}|_{\text{WHERE CustomerId = \$rs2.CustId}}\}$

$[\![17]\!] = [\![12]\!] \cup [\![14]\!]$

4.2 Watermarking of Invariant Parts

In this phase, we may use any of the existing watermarking techniques [15] to watermark the invariant part of the database obtained in the previous phase. As invariant parts are not prone to modification, of course the embedded watermark will behave as persistent one.

However, the choice of existing watermarking technique is determined by (i) the use of data in a particular application context, (ii) the size of invariant part which is used as cover, (iii) the type of the cover, etc.

4.3 Watermarking of Applications Using Opaque Predicates

An opaque predicate is a predicate whose truth value is known a priori [8]. Moden et al. [22] first used opaque predicates in softwares watermarking by inserting dummy methods guarded by opaque predicates. The key challenge to design opaque predicates is that they should be resilient to various forms of attack-analysis. A variety of techniques such as using number theoretic results, pointer aliases, and concurrency have been suggested for the construction of opaque predicates [8]. In addition, Arboit also suggested a technique for constructing a family of opaque predicates through the use of quadratic residues [3]. Arboit's proposal is to encode

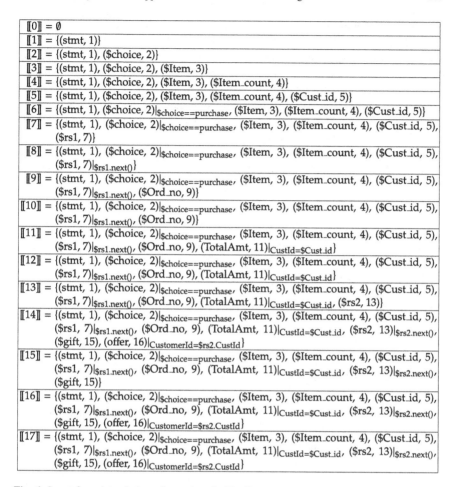

⟦0⟧ = ∅
⟦1⟧ = {(stmt, 1)}
⟦2⟧ = {(stmt, 1), ($choice, 2)}
⟦3⟧ = {(stmt, 1), ($choice, 2), ($Item, 3)}
⟦4⟧ = {(stmt, 1), ($choice, 2), ($Item, 3), ($Item_count, 4)}
⟦5⟧ = {(stmt, 1), ($choice, 2), ($Item, 3), ($Item_count, 4), ($Cust_id, 5)}
⟦6⟧ = {(stmt, 1), ($choice, 2)\|$choice==purchase, ($Item, 3), ($Item_count, 4), ($Cust_id, 5)}
⟦7⟧ = {(stmt, 1), ($choice, 2)\|$choice==purchase, ($Item, 3), ($Item_count, 4), ($Cust_id, 5), ($rs1, 7)}
⟦8⟧ = {(stmt, 1), ($choice, 2)\|$choice==purchase, ($Item, 3), ($Item_count, 4), ($Cust_id, 5), ($rs1, 7)|$rs1.next()}
⟦9⟧ = {(stmt, 1), ($choice, 2)\|$choice==purchase, ($Item, 3), ($Item_count, 4), ($Cust_id, 5), ($rs1, 7)|$rs1.next(), ($Ord_no, 9)}
⟦10⟧ = {(stmt, 1), ($choice, 2)\|$choice==purchase, ($Item, 3), ($Item_count, 4), ($Cust_id, 5), ($rs1, 7)|$rs1.next(), ($Ord_no, 9)}
⟦11⟧ = {(stmt, 1), ($choice, 2)\|$choice==purchase, ($Item, 3), ($Item_count, 4), ($Cust_id, 5), ($rs1, 7)|$rs1.next(), ($Ord_no, 9), (TotalAmt, 11)|CustId=$Cust_id}
⟦12⟧ = {(stmt, 1), ($choice, 2)\|$choice==purchase, ($Item, 3), ($Item_count, 4), ($Cust_id, 5), ($rs1, 7)|$rs1.next(), ($Ord_no, 9), (TotalAmt, 11)|CustId=$Cust_id}
⟦13⟧ = {(stmt, 1), ($choice, 2)\|$choice==purchase, ($Item, 3), ($Item_count, 4), ($Cust_id, 5), ($rs1, 7)|$rs1.next(), ($Ord_no, 9), (TotalAmt, 11)|CustId=$Cust_id, ($rs2, 13)}
⟦14⟧ = {(stmt, 1), ($choice, 2)\|$choice==purchase, ($Item, 3), ($Item_count, 4), ($Cust_id, 5), ($rs1, 7)|$rs1.next(), ($Ord_no, 9), (TotalAmt, 11)|CustId=$Cust_id, ($rs2, 13)|$rs2.next(), ($gift, 15), (offer, 16)|CustomerId=$rs2.CustId}
⟦15⟧ = {(stmt, 1), ($choice, 2)\|$choice==purchase, ($Item, 3), ($Item_count, 4), ($Cust_id, 5), ($rs1, 7)|$rs1.next(), ($Ord_no, 9), (TotalAmt, 11)|CustId=$Cust_id, ($rs2, 13)|$rs2.next(), ($gift, 15)}
⟦16⟧ = {(stmt, 1), ($choice, 2)\|$choice==purchase, ($Item, 3), ($Item_count, 4), ($Cust_id, 5), ($rs1, 7)|$rs1.next(), ($Ord_no, 9), (TotalAmt, 11)|CustId=$Cust_id, ($rs2, 13)|$rs2.next(), ($gift, 15), (offer, 16)|CustomerId=$rs2.CustId}
⟦17⟧ = {(stmt, 1), ($choice, 2)\|$choice==purchase, ($Item, 3), ($Item_count, 4), ($Cust_id, 5), ($rs1, 7)|$rs1.next(), ($Ord_no, 9), (TotalAmt, 11)|CustId=$Cust_id, ($rs2, 13)|$rs2.next(), ($gift, 15), (offer, 16)|CustomerId=$rs2.CustId}

Fig. 6 Least fix-point solution of equations in Fig. 5

the watermark information in the form of opaque predicates and to embed it into the software without affecting the control-flow structures.

The integrity constraints defined on a database ensure that the attributes under the constraints will have right and proper values in the database. Moreover, database designers also have opportunity to define their own assertions. These constraints which in fact define the properties of attribute-values, can be represented in terms of predicate formulas of first-order logic.

In this phase, we identify integrity constraints or we define assertions as a way to represent the properties of values in the variant part of the database obtained in the phase before. Observe that, although values in the variant part are prone to be updated or deleted, their properties represented by the constraints (integrity constraints or assertions) remain unchanged. Importantly, these constraints act as opaque predicate as their truth value w.r.t. the values in variant part is always true. We follow existing software watermarking techniques [16, 23] to watermark the applications in the information system by using these opaque predicates. As the applications contain SQL statements, we may use the conditional-part (WHERE clause) of SQL statements as cover.

Consider the running example. Consider an integrity constraint defined on the attribute "Age" which says that the age must belong to the range 15–70. This is expressed as:

$$15 \leq Age \leq 70$$

Since the formula is always true, it acts as an opaque predicate. Following Arboit's proposal [3], we can watermark the code by embedding this opaque predicate in the statement 13 as shown below:

$rs2 = SELECT CustId FROM Cust WHERE TotalAmt > 5000 AND $15 \leq Age \leq 70$;

5   Complexity Analysis

Let n be the program size. Let p be the number of variables (which include database attributes and application variables) in the program. The number of data-flow equations associated with control-flow nodes of the program is n. Since each data-flow equation depends on the results of the predecessor nodes, the worst-case time complexity of each data-flow equation is $O(n)$. At each iteration the analysis provides us the information about the data defined up to each program point. Therefore, the height of the corresponding finite lattice is $O(p)$. Thus, the overall worst-case time complexity of data-flow analysis is $O(n \times n \times p) = O(n^2 p)$.

6   Security Analysis

The proposed approach focuses on information systems scenario where databases are associated with a predefined set of applications. Our basic assumption is that only the database statements in the associated applications are authorized to perform computations on the database. Since attackers are not allowed to issue any other database operations, this mitigates the possibility of random value modification attacks on watermark in invariant part. This is to note that attacker can perform attacks in the variant part (see in Sect. 7). The integrity constraints, which are treated as opaque predicates, also do not change over time. Therefore, watermark detection in our approach is deterministic in practice. However, attackers may perform static analysis to detect opaque predicates [9] in order to remove watermarks from the associated applications codes.

7 Experimental Results

We have performed experiment on the Forest Cover Type data set.[4] The data set has 581012 tuples and 61 attributes. An extra attribute *id* is added in our experiment that serves as primary key. The experiment is performed on server equipped with Intel Xeon processor, 64 GB RAM, 3.07 GHz clock speed and Linux operating system. The algorithms are implemented in java version 1.7 and MYSQL version 5.1.73.

In Table 2, we describe the notations used in the tables showing experimental results. Table 3 depicts results of watermark detection after random update attacks take place in AHK algorithm [1]. Observe that detection may fail when more tuples are modified (updated) by attackers.

Experimental results obtained in our proposed scheme are depicted in Table 4. We have taken results by changing the size of invariant part as 25, 50, 75 and 90 % that include 145253, 290506, 435759 and 522910 tuples, respectively. Observed that we follow AHK algorithm to embed and detect watermark in invariant part. The experimental results depict that attackers may try to create a new watermark in variant part by performing random modification attacks. The results imply that probability of false-watermark detection in variant part increases if the size of variant part decreases or the value of α (hence τ) decreases. For lower value

Table 2 Descriptions of the notations

Count	No. of tuples used for particular experiment
v	No. of attributes used for marking and detection in the relation
γ	Fraction of tuples used in the experiment
ξ	No. of least significant bit available for marking in an attribute
TC	Total count that is marked during embedding
α	Significance level of the test for detecting watermark
τ	Threshold parameter for detecting watermark

---

[4]Available in the University of California-Irvine KDD Archive kdd.ics.uci.edu/databases/covertype/covertype.html.

Table 3 Detection results after random update attacks in AHK algorithm [1]

Count	ν	γ	ξ	TC	Embed time (msec)	ξ-updated	% tuples updated	α	Match count	τ	Detect time (msec)	Detect?
581012	10	10	15	58166	12058137	10	50	0.9	48432	52349	11874626	×
								0.8		46532	11632480	✓
							70	0.9	44632	52349	11761642	×
								0.75		43624	11494000	✓
			12	58166	11886111	10	50	0.9	53284	52349	11857304	✓
							70	0.9	51376	52349	11219001	×
								0.75		43624	11764361	✓
						8	50	0.9	51499	52349	11523694	×
								0.8		46532	11361032	✓
							70	0.9	49772	52349	11240122	×
								0.75		43624	12085957	✓
581012	10	20	15	28942	12040404	8	50	0.9	27013	26047	11990959	×
								0.8		23153	12058816	✓
								0.7		20259	12321103	✓
							70	0.9	26433	26047	12394587	×
								0.75		21706	11997753	✓
			12	28942	11350339	10	50	0.9	28942	26047	12926076	✓
							70	0.9	28942	26047	12198521	×
						8	50	0.9	24176	26047	12305839	×
								0.8		23153	11789865	✓
							70	0.9	22230	26047	11669565	×
								0.75		21706	11605415	✓

(continued)

Table 3 (continued)

Count	ν	γ	ξ	TC	Embed time (msec)	ξ-updated	% tuples updated	α	Match count	τ	Detect time (msec)	Detect?
581012	10	50	15	11851	12358971	10	50	0.9	9867	10665	11519875	×
								0.8		9480	11358507	✓
							70	0.9	9066	10665	12006882	×
								0.75		8888	11794182	✓
			12	11851	11122883	10	50	0.9	10862	10665	11641533	✓
							70	0.9	10455	10665	11951827	×
								0.75		8888	11689275	✓
						8	50	0.9	10485	10665	11951827	×
								0.8		9480	11983700	✓
							70	0.9	10170	10665	12295857	×

Table 4 Detection after random update attacks on variant in proposed scheme

Invariant part						Variant part							
Count	ν	ξ	γ	TC	Embed time (msec)	Count	ξ-updated	% tuples updated	α	Match count	τ	Detect time (msec)	Detect?
145253	10	12	5	29020	781581	435759	10	50	0.75	43664	65436	10276183	×
			10	14564	781281					21729	32701	9974971	×
			20	7272	781278					10736	16252	9842310	×
			50	2944	780517					4401	6680	9876861	×
290506	10	15	5	58017	2923709	290506	8	50	0.75	29247	43688	7226456	×
									0.5	29247	29125	7207290	✓
								90	0.5	29150	29125	7273410	✓
			10	29046	3128455			50	0.75	14567	21840	7153050	×
									0.5	14567	14560	7037200	✓
								90	0.5	14512	14560	7103890	×
			20	14436	3181587			50	0.75	7241	10879	7313342	×
									0.5	7241	7253	7100858	×
								90	0.5	7222	7253	7329082	×
			50	5889	2975901			25	0.9	2943	5365	7225908	×
								50	0.5	2944	2981	7100437	×
								90	0.5	2922	2981	7353554	×
435759	10	15	5	87202	6653587	145253	8	50	0.5	14572	14533	3993870	✓
								90		14553		4069217	✓
			10	43662	6556017			50	0.5	7269	7252	4017680	✓
								90		7249		3900057	×
			20	21701	6619247			50	0.5	3629	3620	4115240	✓
								90		3625		3938437	✓
			50	8866	6710785			50	0.5	1476	1492	3878820	×
								90		1471		3957758	×

(continued)

Table 4 (continued)

Invariant part						Variant part							
Count	v	ξ	γ	TC	Embed time (msec)	Count	ξ-updated	% tuples updated	α	Match count	τ	Detect time (msec)	Detect?
522910	10	15	5	104641	9691596	58102	8	50	0.5	5901	5813	1612820	✓
								90		5893		1640135	✓
			10	52389	9862129			50	0.5	2921	2888	1713183	✓
								90		2926		1749594	✓
			20	26079	9689954			50	0.5	1453	1431	1700684	✓
								90		1451		1724654	✓
			50	10630	9511040			50	0.5	610	610	1711352	×
								90		611		1686176	✓

of α, attacker may successfully prove the existence of such false-watermark. Parameters used by the attacker for detecting false-watermark are similar as those used for marking by the owner. This situation may arise during proving the ownership in presence of all concerned people.

8 Conclusions

In this paper, we proposed a persistent watermarking of information systems comprising of a set of applications supported by the database at the back-end. We provided a unified framework by combining software watermarking and database watermarking to watermark the complete system at a time. The proposal identifies both variant and invariant part of the database by applying data-flow analysis to the applications, aiming at making the embedded watermarks persistent. The proposed technique serves as generalized framework which may enhance any of the existing techniques in the literature in terms of persistency. We are now in process of building a prototype tool based on the proposal.

References

1. Agrawal, R., Haas, P.J., Kiernan, J.: Watermarking relational data: framework, algorithms and analysis. VLDB J. **12**(2), 157–169 (2003)
2. Al-Haj, A., Odeh, A.: Robust and blind watermarking of relational database systems. J. Comput. Sci. **4**, 1024–1029 (2008)
3. Arboit, G.: A method for watermarking java programs via opaque predicates. In: Proceedings of the 5th International Conference on Electronic Commerce Research (ICECR-5). pp. 184–196. ACM Press, San Diego (2002)
4. Bhattacharya, S., Cortesi, A.: A distortion free watermark framework for relational databases. In: Proceedings of the 4th International Conference on Software and Data Technologies, Sofia (2009)
5. Bhattacharya, S., Cortesi, A.: A generic distortion free watermarking technique for relational databases. In: Proceedings of the Fifth International Conference on Information Systems Security (ICISS 2009). LNCS Springer Verlag, Kolkata (2009)
6. Bhattacharya, S., Cortesi, A.: Distortion-free authentication watermarking. In: Cordeiro, J., Virvou, M., Shishkov, B. (eds.) Software and Data Technologies, pp. 205–219. Springer CCIS, Volume 170 (2013)
7. Camara, L., Li, J., Li, R., Xie, W.: Distortion-free watermarking approach for relational database integrity checking. Mathematical Problems in Engineering (2014)
8. Collberg, C., Thomborson, C., Low, D.: Manufacturing cheap, resilient, and stealthy opaque constructs. In: Proceedings of the 25th ACM SIGPLAN-SIGACT Symposium on Principles of Programming Languages (POPL'98). pp. 184–196. ACM Press, San Diego (1998)
9. Dalla Preda, M., Madou, M., De Bosschere, K., Giacobazzi, R.: Opaque predicates detection by abstract interpretation. In: Johnson, M., Vene, V. (eds.) Algebraic Methodology and Software Technology, pp. 8–95. Springer LNCS 4019 (2006)
10. Guo, H., Li, Y., Liua, A., Jajodia, S.: A fragile watermarking scheme for detecting malicious modifications of database relations. Inf. Sci. **176**, 1350–1378 (2006)

11. Gupta, G., Pieprzyk, J.: Database relation watermarking resilient against secondary watermarking attacks. In: Proceedings of the Fifth International Conference on Information Systems Security (ICISS 2009). pp. 222–236. LNCS Springer Verlag, Kolkata (2009)
12. Halder, R., Cortesi, A.: Persistent watermarking of relational databases. In: Proceedings of the IEEE International Conference on Advances in Communication, Network, and Computing (CNC'10). IEEE CS, India (2010)
13. Halder, R., Cortesi, A.: A persistent public watermarking of relational databases. In: Proceedings of the 6th International Conference on Information Systems Security (ICISS'10). pp. 216–230. Springer LNCS 6503, India (2010)
14. Halder, R., Cortesi, A.: Abstract interpretation of database query languages. Comput. Lang. Syst. Struct. **38**, 123–157 (2012)
15. Halder, R., Pal, S., Cortesi, A.: Watermarking techniques for relational databases: survey, classification and comparison. J. Univ. Comput. Sci. **16**(21), 3164–3190 (2010)
16. Hamilton, J., Danicic, S.: A survey of static software watermarking. In: 2011 World Congress on Internet Security (WorldCIS'11). pp. 100–107. IEEE (2011)
17. Khan, A., Husain, S.A.: A fragile zero watermarking scheme to detect and characterize malicious modifications in database relations. Sci. World J. (2013)
18. Khanduja, V., Chakraverty, S., Verma, O.P., Singh, N.: A scheme for robust biometric watermarking in web databases for ownership proof with identification. In: Active Media Technology, pp. 212–225. Springer (2014)
19. Khanduja, V., Verma, O.P., Chakraverty, S.: Watermarking relational databases using bacterial foraging algorithm. Multimed. Tools Appl. pp. 1–27 (2013)
20. Li, Y., Deng, R.H.: Publicly verifiable ownership protection for relational databases. In: Proceedings of the 2006 ACM Symposium on Information, computer and communications security (ASIACCS'06). pp. 78–89. ACM, Taipei (2006)
21. Li, Y., Guo, H., Jajodia, S.: Tamper detection and localization for categorical data using fragile watermarks. In: Proceedings of the 4th ACM workshop on Digital rights management (DRM'04). pp. 73–82. ACM Press, Washington DC (2004)
22. Monden, A., Iida, H., Matsumoto, K.i., Inoue, K., Torii, K.: A practical method for watermarking java programs. In: Proceedings of the 24th Annual International Computer Software and Applications Conference, (COMPSAC 2000). pp. 191–197. IEEE (2000)
23. Myles, G., Collberg, C.: Software watermarking via opaque predicates: implementation, analysis, and attacks **6**(2), 155–171 (2006)
24. Nielson, F., Nielson, H.R., Hankin, C.: Principles of program analysis. Springer, New York (1999)
25. Sion, R., Atallah, M., Prabhakar, S.: Rights protection for categorical data. IEEE Trans. Knowl. Data Eng. **17**, 912–926 (2005)
26. Yingjiu, L.: Database watermarking: A systematic view. Springer, Berlin (2007)
27. Zhang, Y., Niu, X., Zhao, D., Li, J., Liu, S.: Relational databases watermark technique based on content characteristic. In: First International Conference on Innovative Computing, Information and Control (ICICIC 2006). IEEE CS, Beijing (2006)
28. Zhou, X., Huang, M., Peng, Z.: An additive-attack-proof watermarking mechanism for databases' copyrights protection using image. In: SAC'07: Proceedings of the 2007 ACM Symposium on Applied Computing. pp. 254–258. Seoul, Korea (2007)

A New Framework for Configuration Management and Compliance Checking for Component-Based Software Development

Manali Chakraborty and Nabendu Chaki

Abstract Component-based software development (CBSD) decreases the time and cost for developing high quality software. However, with CBSD, the maintenance of the software is more difficult, as the whole system consists of several composite components. In this paper, a three-layer framework is proposed toward designing an efficiently configurable component-based system. We also developed an algorithm to identify the primitive and composite components that are related in terms of dependency. This helps managing multiple versions of a system. A smart meter system is considered as a case study. Our algorithm is executed on this component-based system using the semantic effect annotations of Business Process Modeling Notation (BPMN) to validate the results of our algorithm. The success reflects the effectiveness of the proposed algorithm toward identifying the components affected by a change in a simple way.

Keywords CBSD · Configuration management · Compliance · Version management

1 Introduction

Modern software systems become more large and complex, because of their improved performance, efficiency, and better quality. Also, the production costs and time for these systems should be minimized. Thus, the maintenance and modifications of those systems are also becoming more critical [1]. Traditional approaches for software development cannot deliver software in short deadlines and with lower costs. A new paradigm called CBSD is used to develop software with existing,

M. Chakraborty (✉) · N. Chaki
Department of Computer Science and Engineering, University of Calcutta, Kolkata, India
e-mail: manali4mkolkata@gmail.com

N. Chaki
e-mail: nabendu@ieee.org

© Springer India 2016
R. Chaki et al. (eds.), *Advanced Computing and Systems for Security*,
Advances in Intelligent Systems and Computing 396,
DOI 10.1007/978-81-322-2653-6_12

already built and used components. In CBSD, the software systems can be developed by selecting off-the-shelf components from some component repository and then integrating them to build the intended software. The components can be developed by different developers using different languages and technologies [2]. Instead of building every software from scratch, CBSD reuses the components, modifies them to satisfy the requirements and then assembles them. This leads to lower cost, smaller development time and better quality of the software, as the components are already built and tested.

The differences between traditional software development (TSD) approaches and CBSDs are listed in Table 1. In CBSD, the management of different components and their versions is one of the most challenging tasks. To achieve this configuration management is used. Configuration management is the task of managing the configuration of different components in a system so that the system operates seamlessly. For a large and complex system, a systematic use of configuration

Table 1 Difference between TSD and CBSD

Property	TSD	CBSD
Development style	Each software is developed from scratch	Already existing components are assembled to build new software. Reusing of software components are the main theme of CBSD
Life cycle	In TSDS the different activities are, requirement analysis, feasibility study, design, coding, testing, maintenance etc.	Life cycle in CBSD consists of, finding components, selecting those that fit the requirements, adapting them, and replacing them with modified versions
Languages	Programming languages are used to implement the system	Primitive components are implemented using programming languages, and composite components are built using component description languages and architecture description languages
System construction	The system is usually implemented by a group of source code files which can be compiled and linked together to form the final system	System construction is a recursive process, in which, primitive components are used to construct composite components. Both primitive and composite components are used to construct larger composite components
Working team	There are engineering teams, which provide all the functionalities during the life cycle of software and end-users	There are component producer teams developing components; consumer teams developing software reusing components and hybrid teams that are both consumer, producer, and end-users

management is used to maintain the correct operability of the components. The different functions of configuration management are [3]: version management, change management, build management, release management, and workspace management.

Authors of paper [4–6] have discussed the various challenges of configuration management in CBSD. They also suggested that run time configuration is needed for CBSD and proposed a model for it. In [7], authors propose a component-based configuration management model, where the components are the integral logical constituents of the system. The model analyzes the relationship among the components and the configuration management part is dependent on that analysis. A model based on the component system and layered architecture is proposed in [8]. Authors claimed that this layered architecture improves reusability and maintainability of configuration management in CBSD. Another distributive, component-based layered model for configuration management in CBSD is proposed in [9]. The layered architecture makes this model easily adaptive, dynamic to changes and brings down the coupling of the system. In [10], dependency graphs identify different types of dependencies among components and analyze them. The graphs are used to facilitate maintenance by identifying differences, i.e., deviations of a configuration from a functioning reference configuration. Based on the unique features of CBSD, we summarize new requirements of CM for CBSD as follows:

(1) For component-based software development, the first step is to select a component from the existing component database. The owners of the database may update the components periodically. If there are more than one versions of a component between two baselines, then there will be two aspects for version management: either store the older versions in the repository, or replace the older versions by the new version. For the first case, the user can use older versions of a component if they want to. However, for the next situation, users are forced to accept the new versions of the components. In Fig. 1, two versions of component 1 exist between two baselines. If a user wants to use version 3 of component 1, then, it will allow doing so, if the older versions of the component are stored in the repository. Otherwise, it has no choice, but to work with the new version of component 1.

(2) Suppose two composite components cc1 and cc2 are dependent on primitive components pc1, pc3, pc5 and pc2, pc3, pc4, respectively, as in Fig. 2. Let us update primitive components pc2, pc3, pc4, and form a new base line for composite component cc2. Now, composite component cc1 is also dependent on primitive component pc3. So, for cc1 there exist two scenarios:

(i) If older versions of primitive components are replaced by the new versions of those components, then cc1 has to adopt itself with the new version of pc3. And the other primitive components of cc1, such as pc1 and pc5, may also need some up gradation to comply with the new version of pc3. Thus, the modification of one component can lead to modification of several components, which may or may not be directly linked with that component.

Fig. 1 Version management
problem in CBSD

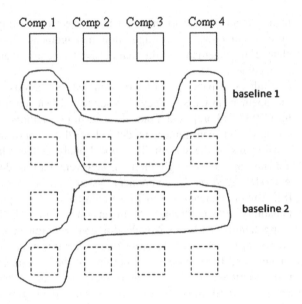

Fig. 2 Dependency between
components

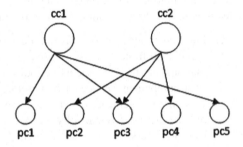

(ii) If older versions of primitive components are kept in a repository, then for a
      given time instant (for a baseline), the two composite components of a system
      will have two different versions of the same primitive components. This may
      lead to a compliance error.

In order to overcome these problems, we propose a new framework for providing configuration and compliance management of components of a system. This
model consists of three layers: component management, configuration management, and compliance management. There exists some research work on configuration management in CBSD, but neither of them incorporates the idea of
compliance management with that. Compliance is a very important factor in CBSD
because the developer imports the components from outside. This makes them
vulnerable for violating the business terms and policies of an organization. Thus,
we propose a layered model which incorporates configuration management and
compliance checking with CBSD. The component management layer deals with the
selection of components and assembling them. It also modifies and replaces

components if necessary. The configuration management layer keeps tracks of the dependencies and relationships between the components. It also analyzes them for achieving maintainability. The compliance management layer is responsible for compliance checking of each component, as well as the total system. We also propose an algorithm to find the effect propagation within a system, when a primitive component is modified. The algorithm is able to identify sets of primitive and composite components, which may get affected with the modification of a certain primitive component. Thus, it helps to configure a system properly.

The smart metering system of smart grid architecture is considered as an application domain for this proposed framework. First, we identify some basic functionalities of smart metering system and consider them as composite components. To achieve each of these functionalities, several small and atomic processes are needed to be executed. We consider all those atomic processes as primitive components. All the composite components, with their respective primitive components are stored in a tabular data structure. When a change request is placed for a primitive component, then our algorithm is used to find how the effects of this change propagate through the system. As an outcome of the algorithm, we can identify the sets of primitive and composite components, which may get affected by the proposed change.

In order to validate the proposed algorithm, Business Process Modeling Notation (BPMN) has been used. BPMN is an agreement between multiple modeling tools' vendors, who had their own notations. BPMN uses a single notation that is understood by all the end-users. It can be used to analyze, simulate, or execute a particular business model. A business process model describes the ordered sequence of different tasks within a process and how the process achieves its objectives [11].

BPMN is an internationally accepted, model-independent tool, which can create a bridge to reduce the gap between business processes and their implementations by providing a unified and standardized graphical representation of any business model. Thus, as a basis of our proposed framework, we use BPMN, to graphically describe smart metering systems, and analyze the effect of component changes on the whole system. When a change request is placed, then the aftereffects of the change for each individual component, as well as the cumulative effect on the whole system is analyzed using semantic effect annotations [12]. This helps identifying the affected components through change propagation and how the system can be reconfigured for a particular change request. The objective is to compare the results produced by the proposed algorithm vis-à-vis finding from BPMN.

An obvious question may arise in this situation: if we can achieve the goal of our algorithm using the semantic effect annotations in BPMN, then why a new algorithm is required at all? Identifying the effects of a change by semantic effect annotation in BPMN needs a certain amount of knowledge about both BPMN and propositional logic. Besides developing the BPMN model for a process and maintaining the semantic effect annotations is a complex task as changing any primitive component results in changing its immediate effect. This change in immediate effect needs to be propagated throughout the system so that the cumulative effects evaluated at different points within the system remain consistent with this change. Also, using BPMN requires maintenance of a graphical representation of the system. On the other hand, our algorithm is inherently simple as it does not require any graphical representations of the system or semantic effect annotations. A simple tabular data structure is sufficient for the execution of the algorithm. As a result, it is a less complex and more preferable solution compared to BPMN.

The rest of the paper is organized as follows: Sect. 2 describes the framework, Sect. 3 explains the functionalities of smart metering system, and the effect of our framework with the help of BPMN and Semantic effect annotation process. We discuss on the future expansion of the work and draw conclusions in Sect. 4.

2 Working Principle of Proposed Framework

In this paper, we proposed a new framework for configuration of components of a component-based software with run time compliance checking. The proposed model has three different parts: (1) component management, (2) configuration management, and (3) compliance management.

2.1 Component Management

The component management part basically deals with the selection of composite and primitive components and maintains their relationships in the form of a list. The functions of the component management are:

Select components	Modify components	Integrate components	Replace components

Select Components The component manager first identifies the composite components of the system. Then for each composite component, primitive components are selected from the component repository.

Modify Components It is not always possible to find the exact component, which meets the requirements of the system. So, then the component manager modifies the components according to the requirements and adapts them to the system.

Integrate Component After collecting all the primitive components, the component manager integrates those primitive components to develop a composite component. The interconnections and dependencies between the composite components are also maintained by the component manager.

Replace Component The component manager also replaces the older versions of a component by the newer and upgraded versions of that component.

The component management layer maintains a data structure for storing the composite components and the primitive components used for each composite

component. Let, there be n composite components, C1 to Cn. For every
Ci ($1 \leq i \leq n$), component manager maintains a list of all of its primitive
components.

Structure *Component_Relation* C
{

    Primitive Component P_1;
    Primitive Component P_2;
    .

    .

    Primitive Component P_n;

}

2.2 Configuration Management

The configuration management part deals with the version management of each
primitive components and how it affects the whole system. Since primitive components are interrelated, modification in one primitive component leads to the
modification of its dependent components. The functions of configuration management are:

Monitor	Select a component for modification	Identify all the related components	Modify	Report to component management	Store

Monitor the configuration manager monitors the whole system to assure that its
working properly and consistently.

Select a component for modification While monitoring the system, the configuration manager also maintains a database for storing the versions of each
component. If a new version of a component arrives in the market, then the configuration manager identifies that component for modification.

Identify all the related components Modifying one primitive component at run
time may affect all the other primitive components related with that component, and
the composite components which are associated with them. So to maintain consistency it is necessary to modify all the other components. Configuration manager
uses an algorithm to identify the related components of an primitive component.

Modify After identifying the components, the configuration manager modifies
the components accordingly.

Report to Component management Then configuration manager reports to the
component manager about these modifications. The component manager then

checks the newly modified components and sends them to compliance manager to make sure that they comply with the business rules of the company.

Store after the compliance checking of the modified components, the configuration manager stores the new versions of those components in a database.

Suppose a primitive component P_j has been modified and a new version of P_j, i.e., $P_j.1$ is introduced. The purpose of this algorithm is to identify the related primitive as well as composite components.

```
Algorithm: Dependency_Analysis(C, Pj, N)
Input: C: Structure Component_Relation C.
       Pj: Primitive component which has been modified.
       N: total number of Composite Components.
Output: PArray - Array for storing primitive components, which are related to a
        particular primitive component Pi.
        CArray - Array for storing the Composite Components, which are related with a
        particular Primitive Component Pi.
{
    for i: = 1 to N
    {
        traverse Ci;
        if (Pj is in the list of Ci and Ci is not in CArray)
        {
            CArray: = Ci;
            for (the rest of primitive components in the
            list of Ci)
                if (they already exist in the PArray)
                    break;
                else
                    put the primate component in PArray;
        }
    }
        end;
}
```

Let us assume that a system has eight composite components. Figure 3 describes the structure for eight composite components. Suppose primitive component P5 has been modified due to some reasons. Therefore, a new version of P5 is introduced as P5.1. In order to maintain the concurrency and compatibility, we must check the other primitive components that are related to P5. In cascade, the composite components which depends on those primitive components will also be checked.

First, we find P5 from the component table. It has been found in the list of C1. Then C1 is added in the C_{Array}, and all other primitive components of C1, i.e., P1 and P2 are added in the P_{Array}. Next, P5 is also in the list of C3. So we put C3 in C_{Array} and P6 in P_{Array}. P5 is not connected with any other composite component. So we take the second element from the P_{Array}, i.e., P1, and repeat the same procedure. P1 is not connected with any other composite component, so we move on to the next primitive component in P_{Array}, P2. P2 is in the list of C4 and C7. So we put both of them in C_{Array}, and add their primitive components, i.e., P8 and P4 in P_{Array}. The next primitive components in P_{Array} are P6, P8, and P4. Since they are not in the list of any other composite components, the procedure is terminated. Figure 4 shows the content of P_{Array} and C_{Array}.

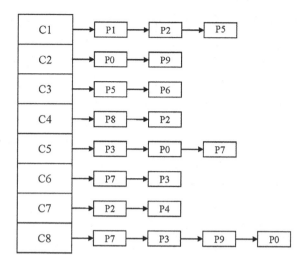

Fig. 3 Component relation structure

P5	P1	P2	P6	P8	P4

C1	C3	C5	C7

Fig. 4 Contents of P_{Array} and C_{Array}

2.3 Compliance Management

The compliance management layer is responsible for checking the compliance of each individual component and as well as the whole system. It also records new compliance rules through the development process of the system. The functions of the compliance management part are:

Check primitive components for compliance	Record new compliance rules	Monitor the integrated system for compliance

Check primitive components for compliance When component manager imports the primitive components from outside, then the compliance manager checks every primitive component for compliance. If they do not comply, then the compliance manager reports to the component management layer, and the component management layer modifies that component accordingly, so that it can comply with the system.

Record new rules While integrating the primitive components, it is sometimes necessary that the components, both primitive and composite, should comply with

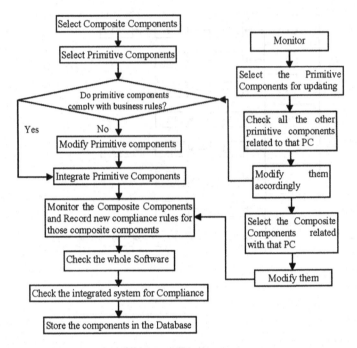

Fig. 5 Workflow diagram of proposed framework

some new rules for successful execution. Thus, another function of compliance manager is to record new compliance rules at run time and keep the business rule database up to date with each change in the system.

Monitor the integrated system for compliance Compliance of each primitive and composite component does not always imply that the whole system is also compliant with the business rules. Therefore, after checking the primitive and composite components for compliance, the compliance manager monitors the whole integrated system for compliance. A baseline is approved only when the system is compliant with the business policies.

The workflow model of the proposed framework is depicted in Fig. 5.

3   Application of this Model in Smart Grid Architecture

In this paper, we consider the smart meter communication architecture of smart grids as an application domain of our proposed model. A smart grid is an intelligent electricity network that integrates the actions of all users connected to it and makes use of advanced information, control, and communication technologies to save energy, reduce cost and increase reliability, and transparency [13].

Smart meter is an advanced energy meter that measures the energy consumption of a consumer and provides added information to the utility company as compared to a regular energy meter [14]. The smart meter communication architecture typically consists of four different components: smart meter, smart energy utility network, DCU (data collection unit), and MDMS (Meter data management System) [15].

The entire scenario is modeled using BPMN. When a change request is placed, then the after effects of the change for each individual component, as well as the cumulative effect on the whole system is analyzed using semantic effect annotations of BPMN. This helps identifying the components affected by the change and how the system can be reconfigured for a particular change request.

We assume that, there are five basic services provided by a smart meter. We consider these five services as five composite components and each composite component further decomposed into several primitive components.

Table 2 provides a detail list of the entire composite and their primitive components for a smart metering system. Now we apply the proposed algorithm on this system and analyze the effect of changing a primate component on the system.

If, primitive component P5 is modified, then the contents of P_{Array} and C_{Array} will be,

$$P_{Array} = P6, P7, P8, P1, P9, P15, P16, P2, P3, P4, P10, P11, P12, P13, P14.$$
$$C_{Array} = C2, C3, C5, C1, C4.$$

Thus, if P5 is modified, then we have to check all the primitive and composite components to check who also need modification. Figure 4 shows the BPMN diagram of the smart meter system. BPMN provides a graphical diagram of how different objectives can be achieved in a business process, with enough information, so that the process can be analyzed, simulated and executed. There are different elements in BPMN—activities, events, gateways, and connectors. A connector links activities, events and gateways and shows the control flow relation. An event can be a start event (start of the process), end event (end of the process), or an intermediate event, that can either be some messages or a timer or error. An activity or a task is an atomic activity and stands for work to be performed within a process. Gateways determine the branching, forking, merging, and joining of paths [11, 16] (Fig. 6).

Immediate effects can be described as the outcome of execution of an activity. This model requires the designers to provide the immediate effects of each activity. Then, the cumulative effect of each component can be calculated by accumulating the immediate effects [12, 17].

In Fig. 4,

- e1 to e16 are the immediate effect of primitive components P1 to P16, respectively.
- CEC1 to CEC5 are the cumulative effect of composite components C1 to C5, respectively. The arrows toward CEC1 to CEC5 mark the points where the cumulative effects have been calculated.

Table 2 Component structure of smart metering system

Composite components	Primitive components
C1: Generate the total electricity consumption of a user	P1: Decode Receive message from DCU
	P2: Collect the total unit of usage
	P3: Generate the bill
	P4: Send message to DCU
C2: Send SMS, if the consumption unit of a user exceeds its previous bill	P5: check the current unit of usage, with previous bill
	P6: Generate an alert message for excess bill amount
	P7: Generate a intermediate bill
	P8: Send message to the user
C3: Alert user before power cuts	P1: Decode Receive message from DCU
	P8: Send message to the user
	P9: Generate an alert SMS for power cut
C4: Services provided for users, who generate electricity in their own houses	P10: Check the electricity generation of a home
	P11: Draw current from home electricity source
	P12: Draw current from outside electricity source
	P13: Check if, generated electricity is sufficient for the home
	P14: calculate the amount of surplus energy and generate a message
	P4: Send message to DCU
C5: take necessary actions, if DCU reports a power shortage	P1: Decode Receive message from DCU
	P8: Send message to the user
	P15: Generate an alert SMS for power shortage
	P16: cut off electricity to some appliances after certain time period

Cumulative effect of C1 (CEC1) = (e1 ∧ e2 ∧ e3 ∧ e4)

Cumulative effect of C2 (CEC2) = (e5 ∧ e6 ∧ e8) ∨ (e5 ∧ e7 ∧ e8)

Cumulative effect of C3 (CEC3) = (e1 ∧ e9 ∧ e8)

Cumulative effect of C4 (CEC4) = (e10 ∧ e12) ∨ (e10 ∧ e11 ∧ e13 ∧ e14 ∧ e4)
                                  ∨ (e10 ∧ e11 ∧ e13 ∧ e12)

Cumulative effect of C5 (CEC5) = (e1 ∧ e15 ∧ e16 ∧ e8) ∨ (e1 ∧ e15 ∧ e8 ∧ e16)

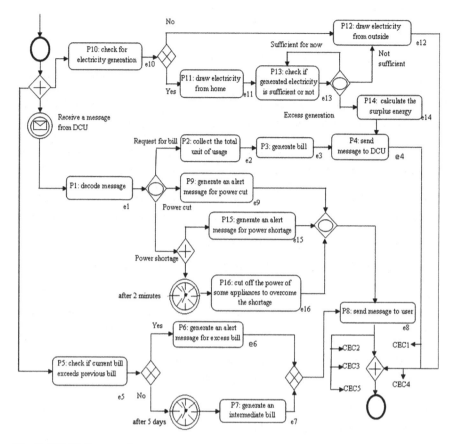

Fig. 6 BPMN diagram of a smart meter

Again, if a change request for P5 is made, then from the diagram and the cumulative effects, we can conclude that,

- Cumulative effect of C2 may get affected, as well as the immediate effect of other primitive components of C2, i.e., e6, e7, e8, and the immediate effects of P6, P7, and P8, respectively.
- Now, P8 is further used in C3 and C5. Hence, if the immediate effect of P8 changes, due to P5, then it may also affect the immediate effects of P1, P9, P15, and P16.
- Again P1 also had contributions in the cumulative effect of C1. Thus, P2, P3, P4 might be affected.
- P4 is also used in C4. So, P10, P11, P12, P13, P14 might also be affected.

Hence, we may conclude that, the BPMN with semantic effect annotation confirms the result of our algorithm.

4   Future Work and Conclusions

One of the most important criteria for a CBSD is to comply with the business policies, rules and regulations of a company. Compliance often refers to the validation of a system against some legal policies, internal policies, or some basic design facts [6]. Compliance checking can be of two types: compliance by detection and compliance by design. In Compliance by detection method, the existing system is checked thoroughly to detect whether it violates any rules or not. If it does not comply, then corrective measures are taken to make it compliant. In compliance by design method, the system is developed, by taking into account the business rules. Thus, the system is designed in such a way, that it can comply with the rules [18]. In CBSD, the system is not developed from scratch. Thus, the compliance by design method does not suit CBSD. Hence, in CBSD, compliance by detection method is used. As an extension of this framework, we would like to work on the detailed working principle of the compliance layer.

In this paper, a new framework for configuration of components with compliance checking is proposed. This framework considers two main problems of CBSD: maintenance and compliance, and solves them by incorporating both configuration management and compliance management with CBSD. In this paper, we consider two-level hierarchy between components, i.e., all the composite components are developed using primitive components. However, the level of hierarchy can easily increased in this model, so that we can consider a scenario where composite components are again assembled together to develop another composite component.

In Sect. 2, requirements for configuration management for CBSD is discussed. Two main problems have been highlighted. One is due to version management, and another is due to the complex and nested relationship between the primitive and composite components. The model in this paper is able to overcome these problems. In this model, the component management layer replaces each component with its latest version once it is accepted by the compliance layer, but the configuration management layer stores all the versions of a component to a database. Thus, the active system is always executed with current versions of each component, but the older versions are also stored in the database.

Also, the component management layer uses a tabular data structure and the configuration management layer uses an efficient algorithm to search all the related primitive and composite components for a particular primitive component. This helps the model to perform efficiently and provides an easily maintainable and compliant system.

Although it is not a theoretical proof for correctness of the proposed algorithm, the validation using BPMN indeed shows the effectiveness of the new algorithm. The proposed methodology builds the foundation for several meaningful extensions in future. We want to apply this model to the entire smart grid architecture as a future work.

5   Acknowledgment

This work is a part of the Ph.D. work of Manali Chakraborty, a Senior Research Fellow of Council of Scientific and Industrial Research (CSIR), Government of India. We would like to acknowledge CSIR, for providing the support required for carrying out the research work.

References

1. Crnkovic, I.: Component-based software engineering—new challenges in software development. J. Comput. Inf. Technol. CIT 11. **3**, 151–161 (2003)
2. Pour, G.: Component-based software development approach: new opportunities and challenges. In: Proceedings Technology of Object-Oriented Languages. TOOLS 26. pp. 375–383 (1998)
3. Estublier, J.: Software configuration management: a roadmap. In: Proceedings of 22nd International Conference on Software Engineering, the Future of Software Engineering. ACM Press, New York (2000)
4. Larsson, M., Crnkovic, I.: Development experiences of a component-based system. In: 7th IEEE International Conference and Workshop on the Engineering of Computer Based Systems ECBS (2000)
5. Larsson, M., Crnkovic, I.: Component configuration management. In Proceedings of ECOOP Conference, Workshop on Component Oriented Programming Nice, France (2000)
6. Lohmann, N.: Compliance by design for artifact-centric business processes. In: 9th International Conference on Business Process Management, pp. 99–115 (2011)
7. Hong, M., Lu, Z., Fuqing, Y.: A component-based software configuration model and its supporting system. J. Comput. Sci. Technol. **17**(4), 432–441 (2002)
8. Mao., M, Jiang, Y.: A new component-based configuration management 3C model and its realization. In: ISISE, International Symposium on Information Science and Engineering, vol. 1, pp. 258–262 (2008)
9. Ruan, L., Yong, Z.: A new configuration management model for software based on distributed components and layered architecture. Parallel Distrib. Comput. Appl. Technol. 665–669 (2003)
10. Larsson, M.: Applying configuration management techniques to component-based systems. Licentiate Thesis Dissertation, Department of Information Technology Uppsala University, vol. 7 (2000)
11. Object Management Group: Business Process Modeling Notation (BPMN) Version 1.0. OMG Final Adopted Specification. Object Management Group (2006)
12. Hinge, K., Ghose, A., Koliadis, G.: Process SEER: a tool for semantic effect annotation of business process models. In: Thirteenth IEEE International Enterprise Distributed Object Computing Conference (EDOC) Los Alamitos, USA, pp. 54–63. IEEE (2009)
13. White Paper by United States Agency for International Development, USAID India: The smart grid vision for India's power sector (2010)
14. Kim, M.: A survey on guaranteeing availability in smart grid communications. Adv. Commun. Technol. (ICACT) 314–317 (2012)
15. Jung, N.J., Yang, K., Park, S.W., Lee, S.Y.: A design of ami protocols for two way communication in K-AMI. In: 11th International Conference on Control, Automation and Systems, pp. 1011–1016 (2011)
16. Goel, N., Shyamasundar, R.K.: An executional framework for BPMN using Orc. APSCC, pp. 29–36. IEEE (2011)

17. Koliadis, G., Vranesevic, A., Bhuiyan, M., Krishna, A., Ghose, A.: Combining i* and BPMN for business process model lifecycle management. In: BPM'06 Proceedings of the 2006 international conference on Business Process Management Workshops, pp. 416–427 (2006)
18. Sackmann, S., Kahmer, M., Gilliot, M., Lowis, L.: A classification model for automating compliance, pp. 79–86. CEC/EEE. IEEE (2008)

CAD-Based Analysis of Power Distribution Network for SOC Design

Ayan Majumder, Moumita Chakraborty, Krishnendu Guha and Amlan Chakrabarti

Abstract Incorporation of power distribution network (PDN) in computer-aided design (CAD) of integrated circuits (ICs) is essential in the recent era. In order to reduce the overall power requirement, the common practice is to reduce the supply voltage. The lowering of supply voltage results in stiffening noise margin and hence increasing the effects of supply voltage fluctuation due to power supply noise. From recent research works, it is also evident that the fluctuation in supply voltage is increasing with scaling down of technology node. A proper estimation of overall power dissipation can only be performed through appropriate and exact parametric extraction of the circuit along with the PDN. Typically there exist many models of voltage fluctuation, which can be utilized to analyze the PDN. In this paper, we propose a new CAD model, which at first estimates a resistance distribution profile of the PDN based on geometric parameters of the chip and electrical parameter of the interconnects (sheet resistance), and then it is mapped with the circuit grid to perform the exact PDN analysis. To the best of our knowledge our proposed model is the first of its kind in regard to PDN analysis. We have chosen one ISCAS 85 benchmark circuit and cryptocores (DES, AES) as SOC applications for our analysis. We have used MATLAB and Mentor Graphics Pyxis tool for our simulation and analysis.

Keywords Power distribution network (PDN) · Cryptocore · Resistance distribution profile · Sheet resistance · Voltage fluctuations

A. Majumder (✉)
Institute of Radio Physics and Electronics, University of Calcutta, Kolkata, India
e-mail: ayan.mdr@gmail.com

M. Chakraborty · K. Guha · A. Chakrabarti
A.K. Choudhury School of Information Technology, University of Calcutta, Kolkata, India
e-mail: moumitachakraborty_it@yahoo.co.in

K. Guha
e-mail: mail2krishnendu@gmail.com

A. Chakrabarti
e-mail: acakcs@caluniv.ac.in

© Springer India 2016
R. Chaki et al. (eds.), *Advanced Computing and Systems for Security*,
Advances in Intelligent Systems and Computing 396,
DOI 10.1007/978-81-322-2653-6_13

189

1 Introduction

The PDN distributes stable power to the circuit components over the entire chip area. PDN is actually a tree, which comprises of branches at each level, viz., motherboard, package, and chip (Fig. 1). Usually the power and ground networks are symmetric. The goal of PDN is to deliver the required current across the chip while maintaining the voltage levels necessary for proper operation of logic circuits. Owing to presence of parasitic elements in PDN, there is a voltage drop in the network. Inductance (L) of the off-chip cables, circuit boards, connectors, package pins, and bond wires and resistance (R) of the on-chip wires cause noise in the supply voltage. The voltage drop across the PDN is commonly referred to as IR drop. In Fig. 1 [1, 2] a simplified PDN circuit is shown.

Different approaches have been investigated to estimate the IR drops (V_{IR}) for chip or package [3]. There are two ways to provide supply currents to the pads of the power grid, either by package leads in wire-bond chips or through C4 (controlled collapsed chip connection) bump-array in flip-chip technology [4, 5]. Both the performance (in terms of speed) of digital circuits, which depends on the parasitic impedance of the real packages, and the internal power supply delivery network can cause fluctuations in the power supply voltages [2, 6]. The delay of digital circuits inversely depends on the supply voltage [7]. As we are moving towards the lower technology nodes with higher device density and faster switching speed, the inductive component is not negligible and becomes comparable to the resistive component [8]. The drop is referred as the LdI/dt drop. Therefore, the effective voltage of the devices is simply described by the following equation:

$$V_{device} = V_{dd} - LdI/dt - V_{IR} \tag{1}$$

Equation 1 shows the effective decrease in device voltage (V_{device}) because of the combined effect of inductive component and IR drop ($LdI/dt + V_{IR}$). Owing to this voltage drop, noise margin is reduced and switching time of gate is increased. The

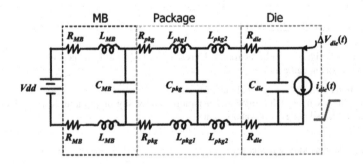

Fig. 1 Simplified model of power distribution network [1, 2]

voltage fluctuations in PDN network can also inject noise in the circuit, which leads to functional failure of the design [1]. Insertion of decoupling capacitances at different levels (on-die, package, motherboard) not only reduces the power supply glitches but also minimizes the ground bounce [9]. But these explicit decoupling capacitors (decap) result in area overhead. Both over estimation and under estimation of decap can degrade the performance of the design. So, estimation of decap is an important concern for in-circuit design [1]. Intrinsic and extrinsic on-die decoupling capacitances may cause power supply resonance by interacting with the package inductance, and that leads to an excessive supply voltage fluctuation in power distribution network. The package and chip have to be viewed as a single network for minimizing resonance.

Figure 1 illustrates the resistance, inductance, and capacitance for motherboard (R_{MB}, L_{MB} and C_{MB}), the resistance and capacitance (R_{pkg}, and C_{pkg}) for package. L_{pkg1} and L_{pkg2} are the different package inductances. All the decoupling capacitors of the die have been considered in a single capacitor C_{die} and a single time-varying current i_{die} is considered to model all the switching devices and R_{die} is the resistance of the die. In PDN, power is distributed through parallel grids of multiple levels of metals [3]. Estimation of total sheet resistance for a chip with multiple grids is described in [3]. Considering this circuit parameter (sheet resistance) and defining the boundary of the conductive foil model, the resistance distribution is illustrated.

The key contributions are summarized as follows:

(i) Practical implementation of an existing mathematical model for PDN extraction.
(ii) A new CAD-based design flow for PDN analysis.
(iii) Incorporation of SOC-based design (benchmark circuits and cryptocores such as DES and AES for SOC applications) for the proposed CAD flow.

In this work, first, a CAD-based design flow is proposed in Sect. 2. In Sect. 3, an optimized resistance distribution profile is obtained using MATLAB. A brief description about the benchmark circuits and cryptocores implemented in this work as a SOC has been explained in Sect. 4. Section 5 observes the power dissipation of the grid in ASIC platform (Mentor Graphics) by considering these benchmark circuits and digital SOC for crypto application. Implementation and results are discussed in Sect. 6. Finally, in Sect. 7, we conclude the paper.

2   Proposed CAD-Based Design

Implementation of the proposed CAD-based design flow (as explained in Fig. 2) is described in the following three steps.

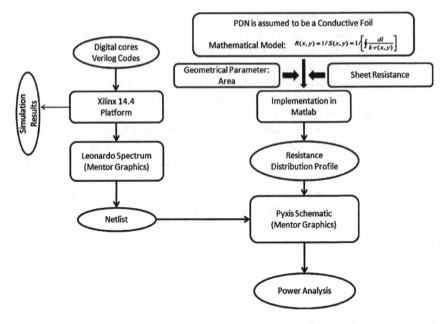

Fig. 2 Proposed CAD flow

2.1 Resistance Distribution Profile Generation in MATLAB

To plot the square profile in MATLAB, the area parameter (xlen, ylen) and sheet resistance ($k = 1$) are declared. Then the perimeter and the boundary of four edge lines: $\{x = 0, y = 0....15\}$, $\{x = 0...15, y = 0\}$, $\{x = 15, y = 0....15\}$, $\{x = 0...15, y = 15\}$ are defined. After that, the integrand of the finite number of points on the boundary is evaluated by using a function. Finally a mean is taken and it is multiplied with perimeter to obtain the desired resistance distribution profile.

2.2 Generation of Circuits and SOC

Gate level modules of circuits and systems are implemented and integrated in ASIC platform. Soft IP modules or the RTL codes of the benchmark circuits and the modules of the SOC are first described in Xilinx 14.4 ISE Platform. Next, RTL simulation is performed in Xilinx ISim Simulator for logic verification. Leonardo Spectrum Platform of Mentor Graphics is used to generate the associated netlist files. Schematic model of the benchmark circuits and SOC is generated by importing the obtained netlist files in the Pyxis Schematic Platform of Mentor Graphics.

2.3   Analysis in ASIC Platform

In the ASIC platform, we used Mentor Graphics Pyxis tool for simulation to verify the correctness of our design. Here we integrate analog circuitry with the digital design like benchmark circuits and cryptocores for our analysis. This helps us to analyze the power with PDN and without PDN for a given technology node.

3   Resistance Distribution Profile

In [6], PDN is considered as a single conductive foil, resistance distribution is obtained by sheet resistance and the dimensions of the chip can be obtained as well.

We assume that the power distribution grid as an isotropic and homogeneous conductive foil with given sheet resistance value represented as k, $k = 1$ in our case (Fig. 3). Using the analytical expression obtained from [6], we evaluate the resistance distribution in the power grid network and then the power consumption at different values of resistance is shown in Table 1. The resistance distribution of the surface of the foil is obtained by considering all the contours that are connected to GND, which requires a single analytical expression [6]. The assumption is followed by the analytical equation:

$$R(x,y) = 1/S(x,y) = 1/\left[\oint \frac{dl}{k \cdot r(x,y)}\right] \qquad (2)$$

where R is the distributed resistive function, S the distribution conductive function, k the sheet resistance and r(x, y) the distance between (x, y), and the perimeter dl, the integration is made around all the perimeter of the chip.

After investigating the square foil, whose sides are arbitrary 15 units long and unitary sheet resistance, the resistance distribution profile that we obtained through MATLAB is shown in Fig. 4.

The maximum value of resistance is 0.4268 Ω.

Fig. 3 Power distribution network as a homogeneous conductive foil characterized by a sheet resistance and a dielectric constant [6]

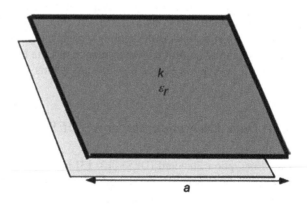

Table 1 PDN power of ISCAS 85 C17 as benchmark circuit in 180-nm technology with resistance distribution profile

Value of resistance applied to grid (Ω)	Without PDN power (pW)	With PDN power (pW)	% increase in power
0.08632	113.6588	113.8154	0.137
0.1064		113.8534	0.171
0.1268		113.8963	0.209
0.1411		113.9230	0.232
0.1624		113.9656	0.269
0.1802		114.0027	0.302
0.2018		114.0459	0.340
0.2283		114.0987	0.387
0.2412		114.1069	0.394
0.2688		114.1798	0.458
0.3015		114.2454	0.516
0.3228		114.2879	0.553
0.3408		114.3235	0.584
0.3626		114.3673	0.623
0.3820		114.4063	0.657
0.4036		114.4495	0.695
0.4268		114.4956	0.736

4 Benchmark Circuits and Associated CryptoCores

Standard circuits and systems, which can be used for reference and testing, are commonly known as benchmark circuits. To analyze the proposed PDN model we use one ISCAS 85 benchmark circuit [10]. However, such circuitry is simple and lacks the required complexity required in practical scenario. Hence, we take refuge in common cryptographic application circuitry. Custom processing hardware design of cryptographic algorithms is commonly known as cryptocores. Cryptocores possess several modules such as key generation module, functionality modules and S-box modules. Data transfer between such modules and a dedicated memory for these cores qualify them as a system on chip (SOC). Such cryptocores [11] are widely used and provide the required complexity for practical scenario. In this work, we refer to hardware implementation of Data Encryption Standard (DES) and Advanced Encryption Standard (AES) algorithms.

4.1 Data Encryption Standard (DES)

Data Encryption Standard (DES) [12, 13] is a 64-bit block cipher, which takes as input a 64-bit plain text and a 56-bit key and generates a 64-bit cipher text as

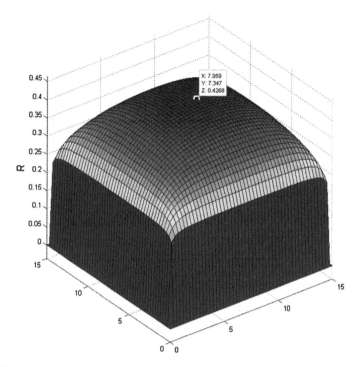

Fig. 4 Resistance distribution profile

output. Its initial and final permutation rounds are predetermined and keyless. A full DES cipher consists of 16 Feistel rounds. Feistel rounds of a practical DES cryptocore are identical. As the number of Feistel rounds increases, the complexity of the cipher increases and proportionally increases the amount of security provided. However, DES of 8 and 12 Feistel rounds can also be used for small, low-security related applications. The round key generator module produces a round key for each Feistel round. Each Feistel round consists of a mixer and a swapper, both of which are invertible functions. The DES function is enclosed inside the mixer, which is the heart of the DES cipher. The DES function comprises of an expansion D box, which increases the input from 32 bits to 48 bits. The XOR function operates on this 48-bit output of the expansion D box with the 48-bit key for the respective round obtained from the round key generator. Eight S boxes are present which provide the real confusion of the cipher. Finally a straight D produces a predetermined permutation. Round 16 is slightly different from the other rounds as it comprises only the mixer but not the swapper.

4.2 *Advanced Encryption Standard (AES)*

Advanced Encryption Standard (AES) [12, 13] is a 128-bit block non-Feistel cipher which takes as input a 128-bit plain text and a 128-bit key and generates a 128-bit cipher text as output. AES implemented in this work comprises of a pre-round transformation along with 10 identical rounds (round 10 is slightly different) and a Key Expansion module, which produces 10 round keys for each round. AES provides four types of transformations namely, Substitute Bytes, Shift Rows, Mix Columns and Add Round Key. The pre-round module uses only Add Round Key transformation, while round 10 does not utilize the Mix Columns transformation. All other rounds use all the four transformations. To create a round key for each round, AES uses a Key Expansion module which creates ten 128-bits round keys for the ten rounds from the 128-bit input cipher key.

5 Analysis in ASIC Platform

The values of resistances, which are obtained from MATLAB implementation, basically applied to the grid (chip) of the power distribution network. The grid consists of resistance, a decap, and also a dc current source. The supply voltage used for analysis is 1.8 V. For our analysis, we consider a benchmark circuit and cryptocores, and connecting them with grid power extraction is done by simulation. In our ASIC platform, 180-nm technology is considered. The value of decap is taken as 39 nF [9]. The values of resistances are obtained from the resistance distribution profile. The block diagram of the incorporation of circuits with PDN is shown in Fig. 5. It describes that PDN exists in both V_{DD} and ground line. For our analysis in ASIC platform our ISCAS benchmark circuits or cryptocores as CUT (circuit under test) are connected with PDN.

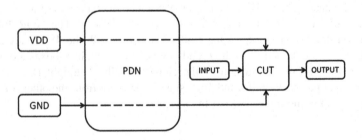

Fig. 5 Incorporation of proposed design model with CUT (circuit under test)

6 Implementation and Results

In this paper, along with the ISCAS benchmark circuit, we have analyzed a DES cryptocore and an AES cryptocore. Varying the number of rounds of a DES cryptocore, gives us the flexibility to analyze the power fluctuations for increasing complexity in practical scenarios. Hence, 8, 12, and 16 Feistel rounds of a DES crypto application are implemented in ASIC platform to give us the desired results. Finally, PDN analysis of an AES cryptocore is performed. Resistance distribution profile has been analyzed in MATLAB. The resistance values have been incorporated into the PDN model in Pyxis Schematic Platform of Mentor Graphics. Design flow of benchmark circuits and cryptocores has been described in Sect. 2.2. We incorporate the digital designs with the PDN circuitry in Pyxis Schematic Platform of Mentor Graphics as described in Sect. 2.3. All simulations in the ASIC Platform have been performed in 180 nm technology node.

In Table 1, we show the PDN power for ISCAS 85 C17 circuit for various resistance values generated through the conductive foil model and mapped-to-grid resistance of the PDN. The range of resistance values given in the first column of Table 1 represents the maximum and minimum values with some intermediate values. We can observe that the PDN power increases with the increase of a resistance value which is expected, and this verifies our model. The incorporation of the PDN in power analysis shows a small percentage increase in total power, which also supports the theoretical PDN model.

In Tables 2 and 3 we show the PDN analysis of SOC circuits DES and AES, respectively. In both the cases, we use the maximum surface resistance value obtained from resistance distribution profile and we find a small percentage increase in total power, i.e., 0.005 and 0.0018 % for DES and AES, respectively. This is not a significant overhead for power.

Table 2 PDN power with the increasing complexity of Data Encryption Standard (DES) circuit in 180 nm technology with the maximum value of resistance

Value of resistance applied to grid (Ω)	DES circuit	Without PDN power (mW)	With PDN power (mW)	% increase in power
0.4268	8 rounds	1.6898	1.6899	0.005
	12 rounds	1.8870	1.8871	0.005
	16 rounds	1.99815	1.99825	0.005

Table 3 PDN power with the increasing complexity of Advanced Encryption Standard (AES) circuit in 180-nm technology with the maximum value of resistance

Value of resistance applied to grid (Ω)	Without PDN power (mW)	With PDN power (mW)	% increase in power
0.4268	5.31345	5.31355	0.0018

7 Conclusion

A PDN binds the voltage drop across the various components of a chip. PDN considered in this scenario is a conducting foil whose mathematical model exists. We implemented a CAD version of such a model by taking the parasitics into consideration. For our analysis, we choose the C17 circuit of ISCAS 85 benchmark series and for a more practical approach, we considered DES and AES cryptocores as SOC applications. Maximum variation in power after inclusion of PDN circuitry for ISCAS C17 benchmark circuit is 0.736 % for various grid resistance values. Next we have considered practical crypto SOC applications. Considering various Feistel rounds of a DES cipher, 0.005 % increment of power is obtained. The maximum percentage increment in power is 0.0018 % when AES cryptocore is considered. Our results confirm that our proposed PDN model is able to stabilize the fluctuations in power with the increase in complexity in practical scenarios. Low power overhead establishes the efficacy of our proposed CAD model of PDN as a conducting foil.

References

1. Pant, S.: Design and Analysis of Power Distribution networks in VLSI Circuits. Thesis report, The University of Michigan (2008)
2. Chakraborty, M., Guha, K., Chakrabarti, A., Saha, D.: Analysis of power distribution network for some cryptocores. In: Proceedings of International Conference on Advances in Computing, Communications and Informatics (ICACCI, 2014), pp 2618–2622 (2014)
3. Shakeri, K., Meindl, J.D.: Compact physical IR-drop models for chip/package co-design of gigascale integration. IEEE Trans. Electron. Dev. 52(6), 1087–1096 (2005)
4. DeHaven, K., Dietz, J.: Controlled collapse chip connection (C4)-an enabling technology. In: Proceedings of Electronic Components and Technology Conference, pp. 1–6 (1994)
5. Turnmula, R.R., Rymwzewski, E.J.: Microelectronics Packaging Handbook, pp. 366–391. Van Nostrand Reinhold, NewYork ch. 6 (1989)
6. Andrade, D., Martorell, F., Moll, F., Rubio. A.: Voltage fluctuations. In: IC Power Supply Distribution Networks: Impact On Digital Processing Systems
7. Rabaey, J.M., Chandrakasan, A., Nikolic. B.: Digital Integrated Circuits a Design Perspective, 2nd edn. ISBN: 0130909963
8. Chen, H.H., Neely, J.S.: Interconnect and circuit modeling techniques for full-chip power supply noise analysis. In: IEEE Trans. Compon. Packag. Manuf. 21(3), 209–215 (1998)
9. Ajami, A., Banerjee, K., Pedram, M.: Scaling analysis of on-chip power grid voltage variations in nanometerscale ULSI. Analog Integr. Circ. Sig. Process 42, 277–290 (2005)
10. https://filebox.ece.vt.edu/~mhsiao/iscas85.html
11. Guha, K., Sahani, R.R., Chakraborty, M., Chakrabarti, A.: Analysis of secret key revealing trojan using path delay analysis for some cryptocores. In: Proceedings of the 3rd International Conference on Frontiers of Intelligent Computing: Theory and Applications (FICTA) 2014, Advances in Intelligent Systems and Computing, vol. 328, pp 13–20 (2015)
12. Stallings, W.: Cryptography and Network Security, 3rd edn. Prentice Hall (2003). ISBN: 0-13-11 1502-2
13. Forouzan, B., Mukhopadhyay, D.: Cryptography and Network Security, 2nd edn. Mc Graw Hill. ISBN: 978-0-07-070208-0

Part III
Algorithms

A New Hybrid Mutual Exclusion Algorithm in the Absence of Majority Consensus

Sukhendu Kanrar, Samiran Chattopadhyay and Nabendu Chaki

Abstract All the voting-based mutual exclusion (ME) algorithms that work on majority consensus inherently confirm safety criterion. However, such algorithms may violate progress condition when no single process gets majority of votes. In this paper, a new two-phase, hybrid ME algorithm is proposed that works even when majority consensus cannot be reached. Simulation results establish that the proposed algorithm offers lower message and time complexity as compared to established as well as recent algorithms. The second phase of the algorithm, in spite of being symmetric, executes in constant time.

Keywords Distributed operating systems · Mutual exclusion · Network partitioning · Majority consensus · Voting · Critical section

1 Motivation and Problem Definition

Mutual exclusion (ME) is crucial for the design of lock-based process synchronization. Depending on the technique used, mutual exclusion algorithms have been classified as token-based [1], time quantum-based [2–4] and permission-based [2, 3, 5–8] algorithms. A classification of different variants of permission-based ME algorithms is presented in Fig. 1.

S. Kanrar (✉) · N. Chaki
University of Calcutta, Block JD, Sector III, Salt Lake, Kolkata 700098, India
e-mail: sukhen2003@gmail.com

N. Chaki
e-mail: nabendu@ieee.org

S. Chattopadhyay
Jadavpur University, Salt Lake, Kolkata 700098, India
e-mail: samiranc@it.jusl.ac.in

© Springer India 2016
R. Chaki et al. (eds.), *Advanced Computing and Systems for Security*,
Advances in Intelligent Systems and Computing 396,
DOI 10.1007/978-81-322-2653-6_14

201

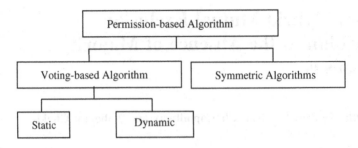

Fig. 1 Permission-based ME algorithms for distributed systems

As this paper primarily deals with hybrid ME that uses a voting mechanism in its first phase, a brief state-of-the-art review on voting-based ME algorithms is presented in Sect. 2.

In voting-based algorithms [9–11], a decision is taken based on voting. Each and every process in a system of n processes need not wait for permission in terms of votes from each of the remaining $n - 1$ processes in the system. Hence, the message complexity of voting-based ME algorithms is lower than the symmetric ME algorithms [8]. There are two types of voting-based algorithms: Static [5] and Dynamic, [12–15] depending on whether the votes assigned remain fixed or not.

The voting schemes [5, 6] are often far from being correct as it may suffer from lack of liveness. Suppose, a network is partitioned and only 85 % of the $n - 1$ processes are connected. If there are two requesting processes, it may lead to a scenario where both the requesting processes get more than 40 % of the votes. However, none of these would achieve majority consensus. A similar situation may occur when there are more than two requesting nodes even if the network is not partitioned and all 100 % of the votes are cast.

Study of the existing literatures has hardly addressed the case where none of the candidate processes achieves majority consensus. In this paper, we have proposed a new hybrid algorithm for ME that finds a candidate for CS when majority consensus is not achieved by any process.

In the event that no single process earns majority of the votes, the classical majority consensus approach [8] suffers from lack of progress. The progress condition demands that at least one of the competing processes must be selected for Critical Section (CS) execution even when majority consensus is not achieved by any process. On the other hand, the liveness property demands that all competing processes must eventually be allowed to enter the CS. This paper aims to address these conflicting requirements while proposing a two-phase hybrid algorithm that ensures both progress condition and liveness even when no single process wins majority vote. The proposed solution also ensures safety conditions so that no two processes enter respective critical sections simultaneously.

Table 1 Classification of permission-based ME algorithms and cited works

Types of permission-based ME algorithms	Reference number
Quorum based or coterie based	[7, 11, 13, 14]
Static voting based	[1, 4, 5, 11, 12, 16, 18, 19]
Dynamic voting based	[6–10, 13, 15–17]
Symmetric or timestamp based	[2, 3]

2 Review on Voting-Based ME Algorithms

As defined above, voting in a ME algorithm can be static or dynamic. Some of the earliest voting-based distributed mutual exclusion algorithms were given by Thomas [5] and Gifford [16]. This was static in nature in which the votes were fixed a priori and the distributed system is assumed to be fully connected by message passing. A node requesting to enter the CS must obtain permission from majority of processes in distributed system. Otherwise, it must not enter the CS and wait until it is allowed to enter the CS.

In Dynamic voting-based approach, a distinguished partition may still have the majority of votes. Dynamic vote reassignment is only possible inside a distinguished partition and the votes of other partitions remain unchanged. This is to avoid data inconsistency as the partitions are unaware of each other. Group consensus and autonomous reassignment are two dynamic vote assignment techniques presented in [17].

Table 1 shows the categorization of different algorithms studied in this paper by putting those in one of the four categories mentioned above.

Timed-Buffer Distributed Voting Algorithm (TB-DVA) [8] is a secure distributed voting protocol. It is unique for fault-tolerance and security as compared to several other distributed voting schemes. TB-DVA is a radical approach to distributed voting because it reversed the two-phase commit protocol: a commit phase (to a timed buffer) is followed by a voting phase. This conceptually simple change has greatly enhanced security by forcing an attacker to compromise a majority of the voters in order to corrupt the system. It may be recalled that in the two-phase commit protocol only one voter must be compromised to corrupt a system. In weighted average algorithm, the weight value is assigned to each process voter input [18], highly available system like distributed database and ad hoc network [12].

3 The Proposed Hybrid Algorithm

The proposed algorithm works in two phases as detailed later in this section. A flow-diagram for the proposed mechanism is presented in Fig. 2.

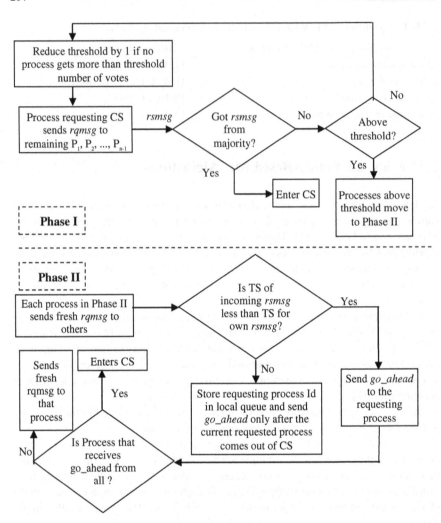

Fig. 2 Flow diagram for the proposed algorithm

Let us state the assumptions required for the proposed algorithm. We consider a distributed system with n processes labeled as N_i for $i \in [1 \ldots n]$:

- Initially, all processes and links are nonfaulty. There is a singular distinguished partition, i.e., a set of processes which elects one candidate.
- A site cannot arbitrary connect to other processes once it has been repaired. Addition is only permitted to a distinguished partition.
- Processes or links may fail before processing the update request in a given site.
- On receiving a request message in phase 1, a process would vote in favor of exactly one process.

- On the contrary, in phase 2, a process may vote in favor of zero or more number of processes.

The proposed algorithm works in two phases. In the event that in the first phase any candidate process receives majority of votes, the algorithm terminates there. This is no different from conventional majority voting algorithm.

However, if no clear winner is identified in phase 1, the second phase is initiated. There is a predefined non-negative integer threshold τ assumed for the proposed algorithm. All the processes that have obtained votes greater than this threshold build a set S and elect the winner in themselves in a second phase of voting. In phase 2, processes that have already earned votes over the threshold τ sends request for a phase 2 vote (P2V). This time with a request for P2V vote, the timestamp (TS) of the original request for CS by the respective process is also sent. Any node $P_Y \in S$ that has received a request for P2V from another node P_Z sends P2V to Z iff $TS(P_Y) > TS(P_Z)$.

It is to be noted here that phase 2 follows a symmetric algorithm to choose the winner. However, as the maximum cardinality of set S cannot exceed n/τ for a total of n competing processes, the overall complexity of the proposed algorithm would be quite low as long as $\tau \gg 0$. It is important toward ensuring both safety and progress condition of the proposed algorithm so that neither two different processes can enter the CS simultaneously nor it leads to a situation where the algorithm comes to a halt without electing a winner.

Procedure Elect-Candidate

Begin

Step 1: All processes wishing to enter CS requests for vote from remaining $n - 1$ processes and waits for the responses.

Step 2: If any process P_i gets majority voting, then go to step 7 else go to step3.

Step 3: If there exists a process that gets vote above a pre-set threshold τ then step 4 else reduce threshold τ by 1 and return to Step 1 to allow all processes to put fresh vote requests.

Step 4: All processes P_X voted above threshold τ are collected in a set S. These processes send request for a phase 2 vote (P2V) along with the timestamp (TS) of the original request from P_X for CS. P_X is to wait for certain predefined time for the responses from all the processes $P_Y \in S$. When time expires, the process Px sends a fresh request for vote and waits for some more time. If no reply is found even after the second attempt, then P_Y treats itself as disconnected process.

Step 5: Any process $P_Y \in S$ that has received a request for P2V from another process P_Z sends P2V to P_Z iff $TS(P_Y) > TS(P_Z)$.

Step 6: Identify processes P_i that receives $(|S| - 1 - n)$ P2V from peers, where n is the number of disconnected processes.

Step 7: Select P_i for entry into its CS.

End.

4 Performance Analysis

In this section, the proposed algorithm is evaluated from multiple perspectives. Issues considered for performance evaluation include correctness of the algorithm in terms of both progress condition and safety, message complexity, fairness, and other important factors of distributed algorithms.

4.1 Safety

A mutual exclusion algorithm satisfies the safety specification of the mutual exclusion problem if it provides mutually exclusive access to the critical section.

Lemma 1 *For the proposed mutual exclusion algorithm, only one process in Phase 2 will get ($|S| - 1 - n$) votes where the set of processes shortlisted for Phase 2 are denoted by S, and n is the total number of processes.*

Proof Without any loss of generalization, one may assume that a standard clock model like Lamport's logical clock model or vector clock would be deployed that puts a unique timestamp for each and every voting request. In Phase 2 of the proposed algorithm, the process $P_i \in S$ which has the smallest timestamp (TS) of its original request for CS would be the winner.

According to the algorithm, the winner is identified when it receives P2V from all other $|S| - 1 - n$ processes short listed for phase 2. Any other process $P_k \in S$ for $k \neq i$ cannot get more than $|S - 2|$ phase 2 votes. This is because (i) P_k will not send itself a vote and (ii) P_i will not send P2V to P_k as TS(P_i) < TS(P_k).

Lemma 2 (Safety) *The proposed voting-based algorithm provides safe mutual exclusion.*

Proof A mutual exclusion algorithm is safe if it ensures that no two processes would enter the respective critical sections simultaneously. In the proposed algorithm, the winner is selected either from phase 1, or from phase 2. The safety is to be considered separately for the two cases.

Case 1: The winner is selected from Phase 1.

This implies that there is a process that gains a majority of votes at the end of phase 1 and hence no other processes can get majority vote. So, only one process is allowed to enter the CS satisfying safety criterion.

Case 2: The winner is selected from Phase 2.

From Lemma 1, only one process would get ($|S| - 1$) number of votes and will be allowed entry to CS and the property of safety is ensured.

Lemma 3 *The value for the threshold integer variable τ as defined in the proposed algorithm would never be negative if it is initialized with some positive integer.*

Proof Let τ be initialized with X, for some $X > 0$. In the event that no process could cross this threshold τ for some iteration of the proposed algorithm, the value of τ is reduced by 1 in Step 3 without checking its present value. We shall show that τ cannot be negative by the method of refutation.

Let us assume that at some point the value of τ becomes -1. This implies (i) in the previous iteration τ was equal to 0 and (ii) no process has obtained any vote greater than $\tau = 0$ in that iteration. But, this is in contradiction with the basic assumption that each process would get at least 1 vote and that from itself. In other words, each process would cross the threshold τ when its value is 0. Therefore, τ would not be reduced further in Step 3. Thus the assumption of $\tau = -1$ is found to be absurd.

4.2 Progress Condition

Progress condition for a mutual exclusion algorithm demands that one of the contending processes for critical section will eventually be allowed to enter the CS, even when no single process gets majority votes.

Lemma 4: (Progress Condition) *Progress Condition is maintained for the proposed mutual exclusion algorithm.*

Proof The proposed algorithm selects a process that has earned majority voting in Phase 1, in case such a process exists. The case where no process gets majority voting is discussed next. From Lemma 3, we conclude that τ cannot be negative and so there are some processes shortlisted for Phase 2. The proposed algorithm always elects at least one process in Phase 2 from the shortlisted one as has been shown in Lemma 1.

Lemma 5 *In spite of being symmetric in nature, the proposed algorithm runs for constant time in phase 2.*

Proof For a threshold of τ, the number of processes that can earn τ votes and get into phase 2 cannot exceed $\eta = n/\tau$. In other words, the cardinality for the set S mentioned in step 4 of the proposed algorithm cannot exceed n/τ. Each of these nodes would send a request for P2V and receives $\eta - 1$ to 0 votes depending up on the timestamp of the request for CS. If threshold is set to as low as 15 % of the total voting processes, then η cannot exceed 6 following the equation $\eta = n/\tau$. This effectively implies a constant time complexity for phase 2 of the execution.

4.3 Correctness

The correctness of control algorithms is typically defined as a collection of safety and liveness. In Sect. 4.1, the safety property has been proved. In Sect. 4.2, progress condition of the algorithm has been proved. Any existing mutual exclusion algorithm [19] that ensures liveness may also be used for this purpose. In this consideration, the proposed two-phase solution provides a framework that is compatible to many different existing voting algorithms that maintain both safety and liveness. The proposed two-phase algorithm therefore holds correctness in tandem with a voting mechanism that is used in phase 1 that ensures liveness in execution.

4.4 Storage Requirement

The proposed solution requires storing very little data at the participating processes. In fact, a process needs to know the total number of processes in the system, and timestamp of its own request for CS.

4.5 Message Complexity

The number of messages per critical section access can be deterministically expressed as a measure of concurrency of requests. Let us assume that a total of m out of n processes want to enter respective critical sections. Each of these m process would request the remaining $n - 1$ processes for vote and a total of $m \times (n - 1)$ requests would be sent. In phase 1, one process is allowed to cast only one vote. Therefore, the number of voting messages would be $(n - 1)$. Thus, in phase 1, the average number of messages exchanged per CS request would be:

$$K = \frac{m \times (n - 1) + (n - 1)}{m} \approx O(n) \tag{1}$$

For a threshold of τ, the number of processes that can earn τ votes and get into phase 2 cannot exceed $\eta = n/\tau$. In other words, the cardinality for the set S mentioned in step 4 of the proposed algorithm cannot exceed n/τ. The symmetric approach followed in phase 2 involves exchange of messages between only these η numbers of processes. Each of these nodes would send a request for P2V and receives $\eta - 1$ to 0 votes depending up on the timestamp of the request for CS. Hence, the average number of messages exchanged in phase 2 for each request to CS would be:

$$\Upsilon = \eta + (\eta - 1) + (\eta - 2) + (\eta - 3) + \cdots + 1$$
$$= \frac{\eta + 1}{2} \tag{2}$$
$$\text{i.e., } \Upsilon \leq n/\tau$$

Therefore, adding the cost from Eqs. 1 and 2, the total number of control messages exchanged is $O(n)$, for a total of n competing processes. Besides, in the proposed algorithm, even when threshold is close to 30 % of the total processes, the number of such processes η entering to phase 2 of the algorithm cannot exceed 3.

If threshold is set to 15 % of the total processes, then η cannot exceed 6 following the equation $\eta = \lceil n/\tau \rceil$. This effectively implies a constant time complexity for phase 2 of the execution. Thus, the algorithm terminates faster and with much lower message complexity compared to what may appear to be the average or even worst case performances from Eqs. 1 and 2.

5 Simulation Result

Similar to our proposed algorithm, the Timed-Buffer Distributed Voting Algorithm (TB-DVA) [8] also uses two-phase commit protocol and Lamport's logical timestamping. Hence, TB-DVA is considered to benchmark the performance of the proposed algorithm in terms of turnaround time for a batch of concurrent processes.

On the other hand, the proposed algorithm, if it has to enter in its phase 2, uses symmetric approach. It is a well-known fact that symmetric algorithms, in spite of being simple to implement, typically involve very high message complexity. Hence, in order to assess message complexity, the proposed algorithm is compared with Ricart–Agrawal's well established symmetric algorithm (RA) [2]. The comparative results of simulation for the proposed method as against RA and TB-DVA is shown in Sect. 5.1.

5.1 Simulation Performance of the Proposed Algorithm with TB-DVA and RA

A connected network topology is considered. The size of the network is gradually increased from 6 to 12 with different connections between the nodes. In order to make a comparative analysis with TB-DVA [8], the value of request time and release time are selected on the same basis as followed for TB-DVA [8]. The difference between the release and the request timestamps is taken as the turnaround time for a particular job. In every case, we compute the average of all results. We also consider the number of request as follows (Table 2):

Network size	6	8	10	12
Number of requests	4	5	6	8

We have considered the length of CS for different jobs as follows:

Site	S1	S2	S3	S4	S5	S6	S7	S8	S9	S10	S11	S12
CS length in time (ms)	3	7	10	2	12	8	11	3	4	7	11	5

Table 2 Simulation parameters

Parameters	Value
Connection topology	Connected graph topology
Number of processes in the graph	Gradually increased from 6–12
Edge length	Static
Request time	Network communication delay
Release time	Twice of network communication delay
Length of CS in terms of execution time	Predefined
Maximum degree of a node	4
Minimum degree of a node	2
Priority of process in the tree	Increases with version number
Maximum number of requests	Gradually increased (4, 6, 8, 10, 16) 95

5.1.1 Time Complexity of the Proposed Algorithm as Compared to TB-DVA

In the proposed algorithm, only the request message is sent to each of the remaining $n - 1$ nodes in the partition. However, for phase 1, the grant messages are considered from only majority of the nodes. Thus, the total number of messages exchanged per token request granted for phase 1 will be $n + \lceil n/2 \rceil$, where n is the total number of nodes in the network. As in [8], the request time is selected according to the network communication delay and release time is considered as twice of the network communication delay. In order to plot the results in graphs, we assume that network communication delay equals to total number of nodes in ms (Fig. 3).

In the proposed algorithm, we have considered 50 % of candidates are granted token in phase 1 algorithm and have calculated the average turnaround time on that basis for every network. Thus, the total turnaround time equals to request time for phase 1 and for phase 2 added with length of CS. The relative performance of the proposed algorithm is better than the TB-DVA algorithm for turnaround time.

In Fig. 4, we consider a fixed size network with 32 nodes and the number of nodes requesting to enter the CS is chosen to be 8(=25 %), 16(=50 %) and 24

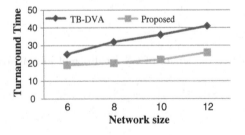

Fig. 3 Turnaround time for TB-DVA and the proposed algorithm

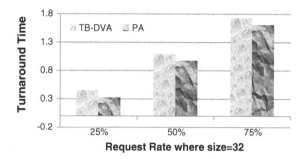

Fig. 4 Comparison between TB-DVA and proposed algorithm for turnaround time

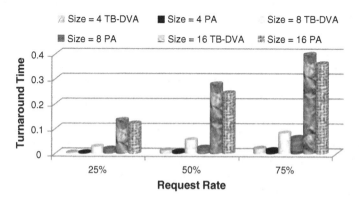

Fig. 5 Execution time of TB-DVA versus PA for increasing network size

(=75 %). In cases, the proposed algorithm works while TB-DVA fails to elect any candidate as no single process gets majority vote. Figure 4 also establishes that the turnaround time of the proposed algorithm is less than TB-DVA in all cases.

Another set of simulation results are generated for networks by gradually increasing its size from 4 to 16. In Fig. 5, we see that proposed algorithm selects a process irrespective of whether majority votes are obtained or not for networks of all sizes. The turnaround time for the proposed algorithm is observed to be less than that of TB-DVA algorithm for all the cases.

5.1.2 Message Complexity of Proposed Algorithm as Compared to RA Algorithm

The simulations setting for comparative performance of the proposed algorithm with analysis with Raymond's algorithm is very similar to that for TB-DVA. However, some variations of these settings such as different number of nodes, etc. have been used in these experiments. These variations are described while explaining respective results.

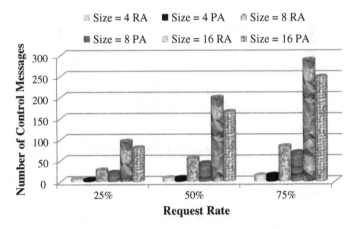

Fig. 6 Control messages for RA algorithm versus proposed algorithm (PA)

In Fig. 6, plots are generated for different network sizes from 4 nodes to 16 nodes. In each of the cases, we consider that 25, 50, and 75 % of nodes have requested to enter the CS. The point to be noted here is that although Ricart–Agrawala's algorithm assumes that the candidate node with lowest timestamp receives consent from all of the remaining nodes, this often does not happen in reality. This is due to message loss or partitioning of the network. In Fig. 6, we observe that the message complexity of the proposed algorithm is less than that of Ricart–Agrawala for all the cases.

The point to be noted here is that although Ricart–Agrawala's algorithm assumes that the candidate node with lowest timestamp receives consent from all of the remaining nodes, this often does not happen in reality. This is due to message loss or partitioning of the network. In Fig. 6, we observe that the message complexity of the proposed algorithm is less than that of Ricart–Agrawala for all the cases.

Algorithm	Evaluation measures				Description
	Message complexity		Synch. delay	Decision theory	
	Heavy load	Light load			
Lamport	$3(N-1)$	$3(N-1)$	T	Static	Prioritize with timestamp
Ricart–Agrawala [2]	$2(N-1)$	$2(N-1)$	T	Dynamic	Get $n-1$ permissions
Quorum dynamic	$O(Q)$	$O(Q)$	3T	Dynamic	Generate dynamic quorum
TB-DVA [8]	$5(N-1)$	$3(N-1)$	2T	two-phase voting	Fault-tolerance and security
Proposed algorithm	$O(N)$	$O(N)$	2T	two-phase voting	No majority consensus needed

(continued)

(continued)

Algorithm	Evaluation measures				Description
	Message complexity		Synch. delay	Decision theory	
	Heavy load	Light load			
Billiard Quorum [20]	$\sqrt{2}\sqrt{N}$	$\sqrt{2}\sqrt{N}$	T	Coterie based	Multidimensional voting
Thomas [5]	$[(N + 1)/2]$	$[(N + 1)/2]$	2T	Majority voting	Introduce the concept of voting

*Q is the number of quorum members

6 Conclusions

Voting-based algorithms for mutual exclusion looking for majority consensus often fails to choose among the candidate processes as none of these may earn majority of votes. In this paper, a new dynamic, hybrid algorithm has been proposed that would work where majority consensus cannot be reached to elect the next process to enter critical section. It is proved that the proposed algorithm maintains progress condition and also ensures safeness and liveness. The solution is compatible to any majority voting-based approach that may be used in phase 1 of the proposed algorithm. The phase 2 essentially selects the candidate process from a group of processes that get votes above a system-defined threshold τ based on timestamp of the original request for entering into the critical section. The solution maintains correctness in tandem with appropriately selected algorithm for voting in phase 1. Simulation results and the theoretical analysis presented in Sect. 5, establish that the message complexity and execution time for the proposed solution is better than the existing solutions that are compared.

References

1. Byeon, B.: NucVoter: a voting algorithm for reliable nucleosome prediction using next-generation sequencing data. ISRN Bioinform. (2013)
2. Ricart, G., Agrawala, A.K.: An optimal algorithm for mutual exclusion in computer networks. Commun. ACM **24**(1), 9–17 (1981)
3. Ekbal, A., Saha, S.: Weighted vote based classifier ensemble selection using genetic algorithm for named entity recognition. NLCS **2010**, 256–267 (2010)
4. Zarafshan, F., Latif-Shabgahi, G.R., Karimi, A.: A novel weighted voting algorithm based on neural networks for fault-tolerant systems. In: Proceedings of the 3rd IEEE International Conference on Computer Science and Information Technology (ICCSIT'10), pp. 135–139 (2010)
5. Thomas, T.H.: A majority consensus approach to concurrency control for multiple copy databases. ACM Trans. Database Syst. **4**(2), 180–209 (1979)

6. Yu, G.X., Glass, E.M., Karonis, N.T., Maltsev, N.: "Knowledge-based voting algorithm for automated protein functional annotation. Proteins Struct Funct Bioinform **61**(4), 907–917 (2005)
7. Saxena, P.C., Rai, J.: A survey of permission-based distributed mutual exclusion algorithms. Comput. Stand Interf **25**(2), 159–181 (2003)
8. Hardekopf, B., Kwiat, K., Upadhyaya, S.: A decentralized voting algorithm for increasing dependability in distributed systems. In: Proceedings of the 7th International Conference on Information System Analysis and Synthesis (ISAS 2001) (2001)
9. Qin, M., Zimmermann, R.: An energy-efficient voting-based clustering algorithm for sensor networks. In: ACIS International Workshop on Self-Assembling Wireless Networks. SNPD/SAWN 2005, pp. 444–451 (2005)
10. Latif-Shabgahi G., Tokhi M.O., Taghvaei M.: Voting with dynamic threshold values for real-time fault-tolerant control systems. In: Proceedings of 16th International Federation of Automatic Control World Congress (IFAC'05) (2005)
11. Hardekopf, B., Kwiat, K., Upadhyaya, S.: Secure and fault-tolerant voting in distributed systems. In: IEEE Aerospace Conference (2001)
12. Zarafshan, F., Karimi, A., Al-Haddad, S.A.R. Saripan, M.I., Subramaniam, S.: A preliminary study on ancestral voting algorithm for availability improvement of mutual exclusion in partitioned distributed systems. In: Proceedings of International Conference on Computers and Computing (ICCC'11), pp. 61–69 (2011)
13. Karimi, A., Zarafshan, F., Al-Haddad, S.A.R., Ramli, A.R.: A novel-input voting algorithm for -by-wire fault-tolerant systems. Sci. World J. 9 (2014). Article ID 672832
14. Latif-Shabgahi, G.R.: A novel algorithm for weighted average voting used in fault tolerant computing systems. Microprocess. Microsyst. **28**(7), 357–361 (2004)
15. Ingols, K., Keidar, I.: Availability study of dynamic voting algorithms. In: Proceedings of the 21st IEEE International Conference on Distributed Computing Systems, pp. 247–254 (2001)
16. Gifford, D.K.: Weighted voting for replicated data. In: Proceedings of the Seventh ACM Symposium on Operating Systems Principles, pp. 150–162. ACM, Pacific Grove (1979)
17. Barbara, D., Garcia-Molina, H., Spauster, A.: Increasing avail-ability under mutual exclusion constraints with dynamic vote reassignment. ACM Trans. Comput. Syst. **7**(4), 394–428 (1989)
18. Azadmanesh, A., Farahani, A., Najjar, L.: Fault tolerant weighted voting algorithms. Int. J. Netw. Security **2**, 240–248 (2008)
19. Osrael, J., Froihofer, L., Chlaupek, N., Goeschka, K.M.: Availability and performance of the adaptive voting replication protocol. In: Proceedings of the 2nd International Conference on Availability, Reliability and Security (ARES'07), pp. 53–60 (2007)
20. Agrawal, D., Egecioglu, O., Abbadi, A.El: Billiard quorums on the grid. Inf. Process. Lett. **64**, 9–16 (1997)

A Comprehensive Sudoku Instance Generator

Arnab Kumar Maji, Sunanda Jana and Rajat Kumar Pal

Abstract Sudoku puzzles have become popular worldwide among many players of different intellectual levels. In this paper, we develop algorithms for creating numerous Sudoku instances of varying levels of difficulty. There are several ways to generate a Sudoku instance. The most popular one is to consider one solved Sudoku puzzle and remove some of the numbers from the cells based on the required difficulty level. Although all Sudoku puzzle creators assume that there is a single solution for a generated puzzle, we notice that a Sudoku puzzle may have multiple solutions. None of the instance generation techniques are able to find how many different solutions are present for a generated Sudoku instance. Here in this paper, we have devised one novel approach that can generate a Sudoku instance and check for its number of possible solutions, and then based on the number of solutions we further categorize the instance that has been generated. This approach is entirely novel and comprehensive for generating Sudoku instances.

Keywords Sudoku · Puzzle · Instance · Difficulty level · Minigrid · Graph theory · Band · Stack

A.K. Maji (✉)
Department of Information Technology, North Eastern Hill University,
Shillong 793022, India
e-mail: arnab.maji@gmail.com

S. Jana
Department of Computer Science and Engineering, Haldia Institute of Technology,
Haldia 721657, India
e-mail: sunanda_jana@yahoo.com

R.K. Pal
Department of Computer Science and Engineering, University of Calcutta,
Kolkata 700098, India
e-mail: pal.rajatk@gmail.com

© Springer India 2016
R. Chaki et al. (eds.), *Advanced Computing and Systems for Security*,
Advances in Intelligent Systems and Computing 396,
DOI 10.1007/978-81-322-2653-6_15

1 Introduction

A Sudoku puzzle can be defined as a partially completed $N \times N$ grid where the initially defined values are known as givens or clues [1]. A general Sudoku puzzle has 9 rows, columns, and minigrids having 9 cells each. So the full grid has 81 cells. Rows, columns, and minigrids are collectively referred to as units or scopes. One rule that can be applied to this puzzle is "Each digit appears once in each unit." For defining the size of the puzzle, often a composite of number of rows × number of columns designation is used, i.e., size 9 × 9. In Fig. 1, a sample Sudoku puzzle is shown along with its possible solution. We can see in this figure that a 9 × 9 Sudoku puzzle is divided into 3 × 3 minigrids. The minigrid is highlighted in bold. For creating an instance of the puzzle as per different levels of difficulty, a proper understanding of the difficulty level is important. So in the next section, we discuss the different difficulty levels of the Sudoku puzzle.

1.1 Metrics of Difficulty Level

In this section, we develop the metrics to determine the difficulty level of a Sudoku puzzle from both computing and human logic deducing perspectives. The following two factors [2] as metrics are taken into consideration, which affect the difficulty level:

- The total amount (or number) of given cells (or clues) and
- the lower bound of given cells in each row, column, and minigrid.

Based on the above two factors with scores, we grade a Sudoku puzzle into five different levels as follows: Level 1: Extremely Easy, Level 2: Easy, Level 3: Medium, Level 4: Difficult, and Level 5: Evil.

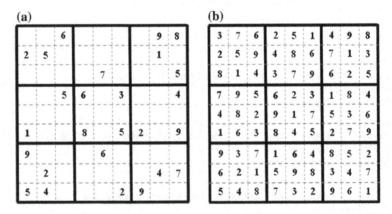

Fig. 1 a An instance of the Sudoku problem. **b** A solution of the Sudoku instance shown in (**a**), where a digit/symbol occurs exactly once in each *row*, *column*, and *minigrid*

Level	Number of clues	Score
1 (Extremely easy)	More than 46	1
2 (Easy)	36–46	2
3 (Medium)	32–35	3
4 (Difficult)	28–31	4
5 (Evil)	17–27	5

Table 1 The amount ranges of givens in each difficulty level

1.1.1 The Total Amount of Given Cells

As the first factor that affects the level estimation is the total amount of given cells in an initial Sudoku puzzle, this factor can significantly eliminate potential choices of digits in each cell by three constraints in the game rule such that each row, column, and minigrid would contain 1 through 9 exactly once. In general, it is reasonable to argue that the more empty cells (and fewer clues) provided at the beginning of a Sudoku game, the higher the level of the puzzle graded in. Different researchers have moderately scaled the amount of ranges of givens for each level of difficulty; in brief such an accumulated data is shown in Table 1 [2].

1.1.2 The Lower Bound on the Number of Clues in Each Row, Column, and Minigrid

The arrangement of empty cells significantly affects the difficulty level if two puzzles provide the same amount of givens (or with slight difference) at the beginning of a Sudoku game. The puzzle with the givens in clusters is graded at a higher level than that with the givens in scattered form. Based on the row, column, and minigrid constraints, we normalize the lower bound on the number of given cells in each row, column, and minigrid for each level of difficulty as shown in Table 2.

1.2 Uniqueness of Sudoku Instances

There is debate among researchers that, if a Sudoku instance has more than one solution, whether this could be considered as a valid Sudoku instance [4]. The

Level	Lower bound on the number of givens in each row, column, and minigrid	Score
1 (Extremely easy)	5	1
2 (Easy)	4	2
3 (Medium)	3	3
4 (Difficult)	2	4
5 (Evil)	0	5

Table 2 The lower bound on the number of clues in each row and column for each difficulty level

definition of Sudoku states that every puzzle must have a unique set of numbers in each row, column, and minigrid.

There is nothing in the rule that states that every puzzle must have a unique solution [3]. Rather, it is the puzzle creators who put an additional constraint that says that a Sudoku instance is valid only if it produces a unique/single solution [4]. However, nobody has effectively explored whether a 'well-built' Sudoku instance (of more than 17 clues) has really one and only one solution [5]. Some mathematicians also debate that Sudoku instances can all be solved by certain standard tricks, many of which may result in a unique *rational* solution to the integer programming problem [5]. We can find rational solutions with linear programming, and if the rational solution is unique, such type of integer programming problem is not NP-hard; it is in P [5]. However, Sudoku is known to be an NP-complete problem [6]. So, it cannot be said that every Sudoku instance must have a unique solution, or otherwise the Sudoku instance is not valid.

We have found several Sudoku instances with more than one solution in the literature [7, 8]. Hence, we can say that a subset of Sudoku instances may have one and only one valid solution, but in general, a Sudoku instance might have two or more solutions as well. This situation has its own merits and applications. Solving Sudoku puzzles has an application in steganography. If a given Sudoku instance has only one valid solution, then that instance may fail to conceal some information while transmitting the instance that an intruder may attack. On the other hand, if there are two or more solutions for a given Sudoku puzzle, then such attackers may be confused in extracting the hidden information.

The most common method for creating a Sudoku instance is the *Digging Hole* strategy. Here, a solved Sudoku grid is taken as input. Then based on the required difficulty levels the values from the different cells are removed on the basis of the constraints provided in Tables 1 and 2. Then it is checked for uniqueness, i.e., whether there exists a unique solution for this puzzle game. There is no methodology that can create Sudoku instances with multiple solutions. We have designed an effective methodology that can check for multiple solutions. Based on the number of solutions, it can categorize the generated instance.

2 Generating a Sudoku Instance from a Solved Sudoku Puzzle

In this method an already solved Sudoku puzzle has been taken as input. Then using *Digging Hole* strategy [9] some of the values (or digits) are removed from the filled-in solved Sudoku grid and then in the end it is checked for conflicts, if any. If there is no more conflict and the instance generated is solvable, then it is taken as a valid Sudoku instance. For example, consider the solved Sudoku puzzle shown in Fig. 2.

Fig. 2 A solved Sudoku
puzzle

1	2	3	4	5	6	7	8	9
6	4	5	7	9	8	3	2	1
9	8	7	3	1	2	5	4	6
8	5	9	6	7	3	4	1	2
4	1	6	2	8	5	9	3	7
7	3	2	9	4	1	8	6	5
5	6	4	8	2	9	1	7	3
3	7	1	5	6	4	2	9	8
2	9	8	1	3	7	6	5	4

Now we can remove the values from any of the cells and check for conflicts, if they arise. *Conflict checking* means to judge whether the generated Sudoku instance provides only one solution after removal of a number of values from the cells (from a solved Sudoku solution in the desired fashion) [9], though before our research we found no existing work that could compute two or more solutions of a Sudoku instance, if it really has. The number of values that are supposed to be there as given clues depend entirely on the difficulty level under consideration, according to the guidelines shown in Tables 1 and 2.

Now we summarize the processes that could be adopted in removing the values from the cells while generating instances from a solved Sudoku puzzle as (1) Randomized and (2) Sequential.

2.1 Randomized Selection of Cell Location

In case of randomized site selection, we can remove the values from any of the locations from a solved Sudoku puzzle. For example, we have removed values from 34 random cell locations from the solved Sudoku puzzle shown in Fig. 2 and obtained a generated Sudoku instance as shown in Fig. 3. Here we have taken away the values from cells [1, 2], [1, 3], [1, 6], [1, 7], [1, 2], [2, 4], [2, 7], [2, 8], [3, 3], [3, 5], [3, 6], [3, 9], [1, 4], [3, 4], [4, 5], [4, 8], [3, 5], [5, 5], [5, 8], [2, 6], [4, 6], [6, 7], [6, 8], [2, 7], [5, 7], [7, 9], [1, 8], [4, 8], [6, 8], [7, 8], [1, 9], [4, 9], [6, 9], and [9, 9].

In Fig. 3, we can say that the generated Sudoku instance is extremely easy as there are 47 givens/clues now and each row, column, and minigrid has at least five givens/clues. Similarly, we can create puzzles with different other levels of difficulty satisfying the constraints mentioned in Tables 1 and 2.

Fig. 3 An extremely easy
Sudoku instance created after
randomly digging holes in the
solved Sudoku puzzle shown
in Fig. 2

1			4	5			8	9
	4	5		9	8			1
9	8		3			5	4	
	5		6		3	4		2
4	1		2		5	9		7
7		2		4	1			5
5		4	8		9	1	7	
	7	1		6			9	8
	9	8		3		6	5	

2.2 Sequential Selection of Cell Location

In this method of generating Sudoku instances, the removal of values from cells can
follow a certain sequence along rows or columns. In case of row-wise direction, the
selection path could be zigzag or like the English letter (capital) 'S' or the reverse
and only row-wise or column-wise or the reverse, as discussed in brief as follows:

2.2.1 Wandering Along S (or Zigzag) Path

In this process, the cells to be dug out are chosen from the left to right directions for
the first row. Then for the next row it starts moving from right to left. Then for the
successive rows, the direction of selection of cells changes alternatively. The
process is shown in Fig. 4a. In a similar way, the direction can also be adopted from
right to left for the first row, from left to right for the second row, and so on, i.e., the
opposite (or reverse) direction of the above.

(a)   **(b)**

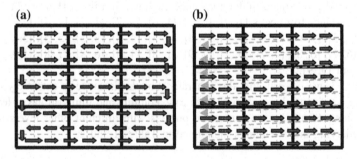

Fig. 4 Cells are dug in *row-wise* direction while **a** wandering along 'S' path and **b** moving from
the *left* to *right* direction

2.2.2 Wandering from Left to Right or the Reverse

In this process, cells to be dug out are chosen from the left to right direction for each of the rows. It starts choosing the cell from the top left corner, then moves to the right, again starts moving from left to right for the next row, and so on. The process is depicted in Fig. 4b. It can also follow the reverse sequence.

Likewise, the cells to be dug can also be chosen in the following fashion when we follow column-wise direction.

(A) Top to Bottom or the Reverse

In this case, the direction of considering the path could be from the top to bottom always for each of the columns, either from left to right or from right to left, or the reverse fashion in all respects.

(B) Wandering along S in Top to Bottom Direction and Vice Versa

In this case, for a solved Sudoku solution, the instance generator may follow a path alternatively from top to bottom and from bottom to top like a sideways 'S' (or zigzag), or the reverse.

Instead of removing the values from the cells randomly or in any well-defined fashion, they can also be removed in a symmetrical way for matching pairs of rows, columns, and minigrids. These are also briefly discussed and exemplified as follows.

2.2.3 Symmetrical Removal of Values from Rows

In this method, values are removed in a symmetrical fashion. For example, if the top row is considered from left to right, then the bottom row is considered from right to left. Then the locations are removed simultaneously and either the corresponding values are both kept at their own locations or both of them are removed. The same manner of considering rows (from the top and the bottom) and keeping or removing values in a symmetrical manner is executed till the middle of the Sudoku instance is reached. The same process can also be executed column-wise in achieving a new Sudoku instance.

For example, a generated instance is shown in Fig. 5, which is created from the solved Sudoku puzzle shown in Fig. 2. This is by definition is a puzzle of medium level difficulty as there remain only 35 clues (or givens) and each row, column, or minigrid contains at least three among the clues. Here we may observe that in the first row (from left to right) the status of clues is ×2××5×78× (including blank cells being represented by '×') and the status of clues in the last row (from right to left) is ×5××3×89× (that are symmetric). Such similarity could be observed for each pair of equidistant rows from the top as well as from the bottom, and also for each pair of equidistant columns from the left as well as from the right in the generated Sudoku instance.

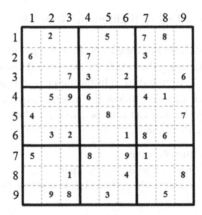

Fig. 5 A Sudoku puzzle instance is created by symmetric removal of values from the solved puzzle shown in Fig. 2. *Row* numbers and *column* numbers are shown on the left and top side of the puzzle

2.2.4 Symmetrical Removal of Values from Columns

If we keenly observe the way of creating symmetrical Sudoku instances, as described in the previous section when we removed (or kept) values row-wise for a solved Sudoku puzzle, it automatically generates instances where it is symmetric column-wise as well. Rather, we may execute the tasks that we followed in generating Sudoku instances by removing (or keeping) values in symmetrical fashion by rows, which can also be obtained by executing this method. For example, we may observe the third column from left (from bottom to top) and the third column from right (from top to bottom); the status of the said columns is 81×2×97×× and 73×4×81××, which are absolutely symmetric.

2.2.5 Symmetrical Removal of Values from Minigrids

In a similar fashion, a new instance of the Sudoku puzzle can also be generated by symmetric removal of values from the minigrids. Based on the symmetry of a Sudoku puzzle, we may observe that if either of the techniques discussed in the above two sections is executed, a Sudoku instance is generated wherein a symmetric fashion could be obtained between pairs of minigrids (1, 9), (2, 8), (3, 7), and (4, 6) in some fashion. We may observe that the created Sudoku instance in Fig. 5 follows these minigrid pairs (in reverse direction). Another minigrid pair could be formed where either column-wise (1, 3), (4, 6), and (7, 9) are symmetric where the values from top to bottom along columns 4 and 6 are also symmetric, or row-wise (1, 7), (2, 8), and (3, 9) are symmetric where the values from left to right along rows 4 and 6 are also symmetric.

2.3 *Flowchart at a Glance for the* **Digging** Hole *Strategy*

The steps for the said algorithm are depicted in the flowchart shown in Fig. 6. From the flowchart, it is clear that first we need to input one solved Sudoku puzzle. Then we have to provide the level of difficulty for which an instance is supposed to be created. From the difficulty level, the algorithm finds out the "can-be-dug" cells (in some fashion, well-defined or arbitrary) based on the constraints and criteria shown in Tables 1 and 2. Then from the "can-be-dug" cells, cells are chosen in some

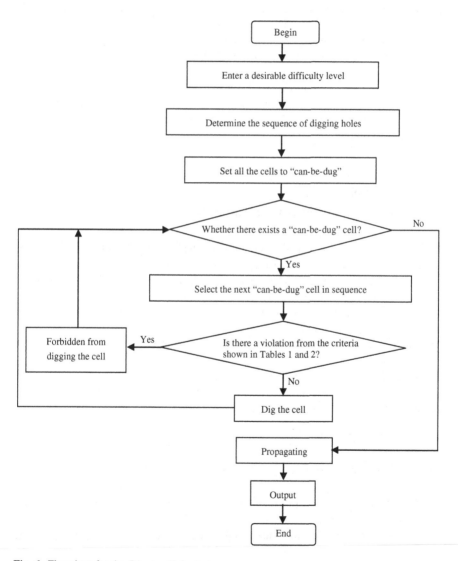

Fig. 6 Flowchart for the *Digging Hole* strategy

fashion and follow a sequence of cells based on the techniques discussed in the previous sections. Next, the chosen cells are made blank by removing values from each of them and checked for if a unique valid solution for the created Sudoku instance is achieved.

Using this method, we can eventually create a Sudoku instance, though this algorithm has not stated clearly from which location(s) the values should be removed. We can remove them from any place we like to, either randomly or by following a definite sequence. Then after removing a value each time, it is checked for whether there exists a unique solution of the generated instance. Again, checking for the solution is a time-consuming task. Incidentally, there is no suitable technique in the literature that can successfully check whether the generated Sudoku instance truly has more than one solution. Another prime limitation is that this strategy needs a solved Sudoku puzzle for generating a Sudoku instance or more.

To remove these limitations, we proposed a new Sudoku instance generation technique, which generates a new Sudoku instance after transformation of the existing instance and then checks for the number of valid solutions based on a graph theoretic formulation.

3   A New Scheme for Creating an Instance Based on Transformations of a Sudoku Puzzle

In this section, we propose a new methodology that creates a Sudoku instance based on the transformation of the existing Sudoku puzzle instance. Then it checks the number of solutions of each of the generated instances based on a graph theoretic formulation [10]. Rather, we may verify the number of solutions possible for a given Sudoku instance from which we would like to compute a large number of such instances using different techniques of transformation.

For transformation, we use methodologies such as (i) *Digit exchanging*, (ii) *Rotation*, (iii) *Rows-in-a-Band exchanging*, (iv) *Columns-in-a-Stack exchanging*, (v) *Band exchanging*, (vi) *Stack exchanging*, and (vii) *The combination of all six methods*. We consider each of these methods and briefly discuss them as follows.

3.1   Digit Exchanging

It is simple to accomplish the method of digit exchanging, because what is necessary here is to exchange all the digits in the cells of one existing Sudoku instance in some well-defined fashion. Interestingly, this exchange does not influence the uniqueness of a Sudoku instance. Thus, a new instance of the Sudoku puzzle is produced.

Figure 7 shows an instance of Sudoku puzzle. Let us replace all 1's belonging to this puzzle by 9 and the reverse. Then the modified Sudoku instance can be as shown in Fig. 8. We can see in this figure, all 9's present in cells [1, 6], [2, 6], and [7, 9] are now replaced by 1's, whereas all 1's present in cells [1], [2, 7], [6, 8], and [9] are now replaced by 9's. All these replacements are highlighted in the figure. This process of exchanging digits is valid throughout the puzzle as a new Sudoku instance is generated, and it is valid for exchanging any number of values among themselves in the puzzle.

Now, the same method can be carried out for multiple pairs of numbers (to be exchanged) as well. A new Sudoku instance can also be produced by replacing values of all clues of a Sudoku instance among themselves. Consider the same Sudoku instance shown in Fig. 7. Here, we may exchange the following values pair wise: (1, 9), (2, 8), (3, 7), (4, 6), (5, 4), (6, 5), (7, 3), (8, 2), and (9, 1). Then, we can get a new instance as shown in Fig. 9. Here we may observe that the 1's present in cells [1], [2, 7], [6, 8], and [9] are now replaced by 9's, the 2's present in cells [1, 9] and [5, 6] are replaced by 8's, and so on that have been depicted in this figure.

Fig. 7 An instance of Sudoku puzzle

1								2
	8				9		3	7
7			5	3			8	
	8			7	3		5	4
		6	4		2	7		
9	7		8	5			1	
	1			8	7			9
3	4		6			8		
8								1

Fig. 8 A new instance is generated from the Sudoku instance shown in Fig. 7 after interchanging of 1 and 9

9								2
	8				1		3	7
7			5	3			8	
	8			7	3		5	4
		6	4		2	7		
1	7		8	5			9	
	9			8	7			1
3	4		6			8		
8								9

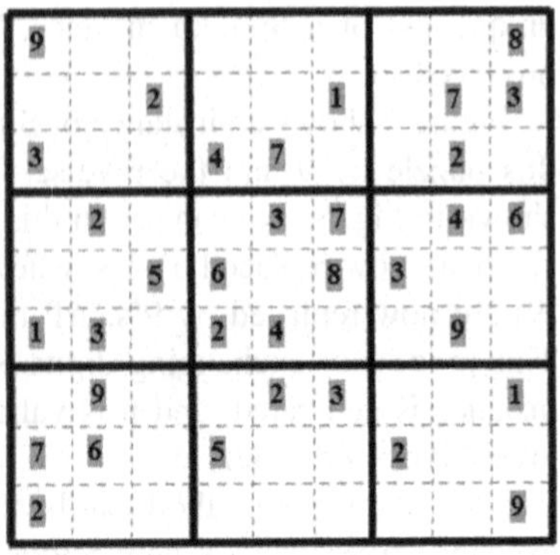

Fig. 9 A new Sudoku instance is generated after replacement of all the digits for the Sudoku instance shown in Fig. 7

3.2 Rotation

In this method, an existing Sudoku instance is rotated by a certain angle (with multiple of unit value 90°) to produce a new instance. By the application of angle of rotation the newly Generated instance can be characterized as follows:

(i) Rotation by 90°, (ii) Rotation by 180°, (iii) Flipping vertical rotation, and (iv) Flipping horizontal rotation.

3.2.1 Rotation by 90 Degree

The Sudoku instance can be rotated by an angle of 90°. The direction of rotation can be of two types: (a) Left Rotation (or in anticlockwise direction) and (b) Right Rotation (or in clockwise direction).

In Fig. 10, a newly generated Sudoku instance is shown, which is produced after left rotation of 90° of the Sudoku instance shown in Fig. 7. We can observe that the clues present in minigrids 3, 6, 9, 2, 5, 8, 1, 4, and 7 in Fig. 7 become the clues of

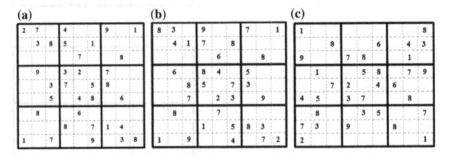

Fig. 10 a A new Sudoku instance generated after left rotation of 90° of the Sudoku instance shown in Fig. 7. **b** A new Sudoku instance is generated after right rotation of 90° of the Sudoku instance of Fig. 7. **c** A new Sudoku instance is generated after rotation of 180° of the Sudoku instance shown in Fig. 7

minigrids 1 through 9, respectively, as shown in Fig. 10a. The row numbers of the clues now become the corresponding column numbers of the same.

After a right rotation of 90° of the same Sudoku instance in Fig. 7, the newly transformed instance is depicted in Fig. 10b. We can observe that clues present in minigrids 7, 4, 1, 8, 5, 2, 9, 6, and 3 become the clues of minigrids 1 through 9, respectively. Here the column numbers of the clues now become the corresponding row numbers of the same.

3.2.2 Rotation by 180 Degree

In the similar way, after 180° rotation of the Sudoku instance shown in Fig. 7, the transformed instance obtained is shown in Fig. 10c. We can observe that the clues present in minigrids 9, 8, 7, 6, 5, 4, 3, 2, and 1 have become the clues of minigrids 1 through 9, respectively, and the initial instance has been toggled that means the top rows (from the left to right) have now been converted to as the bottom rows (from the right to left), and the reverse.

3.2.3 Flipping Vertical Rotation

The Sudoku instance can be flipped vertically (keeping the clues of the fifth row unchanged) to produce a new instance of the puzzle. The Sudoku instance shown in Fig. 11a is produced after flipping vertically the Sudoku instance shown in Fig. 7. We can observe that the clues present in minigrids 7, 8, 9, 4, 5, 6, 1, 2, and 3 have become the clues of minigrids 1 through 9, respectively, as well as the top row of the instance now becomes the bottom one, and vice versa, keeping the row information from the left to right as it was.

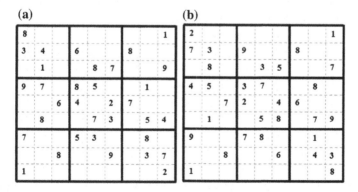

Fig. 11 **a** A new Sudoku instance is generated after flipping vertically f the Sudoku instance shown in Fig. 7. **b** A new Sudoku instance is generated after flipping horizontally of the Sudoku instance shown in Fig. 7

3.2.4 Flipping Horizontal Rotation

An existing Sudoku instance can be flipped horizontally (keeping the clues of the fifth column unchanged) to generate a new instance. In case of flipping horizontal rotation, the clues present in minigrids 3, 2, 1, 6, 5, 4, 9, 8, and 7 now become the clues of minigrids 1 through 9, respectively, as well as the left column of the instance now becomes the right column, and vice versa, keeping the column information from the top to bottom as it was. In other words, each minigrid is now the mirror of the given (or original) one. The transformed minigrid of the Sudoku instance shown in Fig. 7 is depicted in Fig. 11b.

3.3 Rows-in-a-Band Exchanging

Rows-in-a-band exchanging means interchanging two or three rows in the same band in any fashion we like to (or randomly). Three consecutive minigrids in a row form a band as shown in Fig. 12a. We can observe from the figure that minigrids 1, 2, and 3 form band 1, whereas minigrids 4, 5, and 6 form band 2, and minigrids 7, 8, and 9 form band 3. Now we may examine that all the rows in each band are interchangeable among themselves in order to generate newer Sudoku instances.

After the exchange of the rows in the same band, the newly made puzzle obtained is still a valid one. Figure 12b shows a new Sudoku instance, which is created by exchanging the first row with the second row in band 1 for the Sudoku instance shown in Fig. 7. This idea can be generalized in exchanging rows in respective bands, in isolation or in combination, in order to produce more and more Sudoku instances.

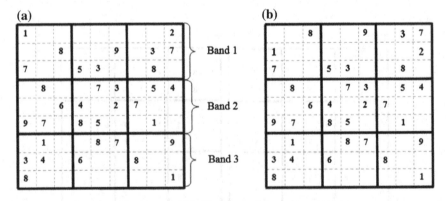

Fig. 12 a Concept of band in Sudoku puzzle, b A new Sudoku instance is created after exchanging the values present in *rows* 1 and 2 in band 1 for the Sudoku instance shown in Fig. 7

3.4   Columns-in-a-Stack Exchanging

Here, we adopt the same concept that has been discussed in the previous section. Columns-in-a-stack exchanging means interchanging of two or three columns in the same stack randomly (or in some well-defined fashion). Let us first understand the concept of stack. The entire minigrids present in the same column form a stack. The concept of stack is shown in Fig. 13a.

From the figure, we can observe that minigrids 1, 4, and 7 form stack 1, minigrids 2, 5, and 8 form stack 2, and minigrids 3, 6, and 9 form stack 3. The columns of each stack are interchangeable among themselves. A newly generated Sudoku instance is depicted in Fig. 13b after exchanging the first two columns of stack 1 of the instance shown in Fig. 7.

3.5   Band Exchanging

Now it is intuitively obvious that we can also interchange the position of an entire band with another one to acquire a new Sudoku instance. Thus, as shown in Fig. 13c, bands 1 and 3 of the Sudoku instance in Fig. 7 have been interchanged to realize this instance. Here we can observe that minigrids 7, 8, and 9 in Fig. 7 have been exchanged with minigrids 1, 2, and 3 to get this new Sudoku instance.

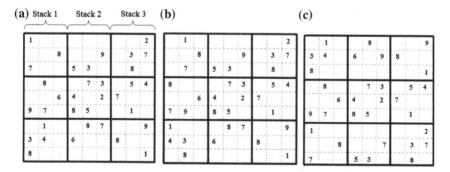

Fig. 13 a Concept of stack in Sudoku puzzle, **b** A new Sudoku instance is created after exchanging the values of *column* 1 and *column* 2 of the Sudoku instance shown in Fig. 7, **c** A new Sudoku instance is created after exchanging the values in band 1 with that of band 3 for the Sudoku instance shown in Fig. 7

3.6 Stack Exchanging

We can also exchange the position of a whole stack with another stack to get a new Sudoku instance. Hence, if stacks 1 and 3 of the Sudoku puzzle shown in Fig. 7 are swapped, and then a new Sudoku instance is obtained as shown in Fig. 14a. We may observe that the clues present in minigrids 1, 4, and 7 are now the clues in minigrids 3, 6, and 9, respectively, and the clues present in minigrids 3, 6, and 9 are now the clues in minigrids 1, 4, and 7, respectively, in the newly generated instance.

3.7 A Combination of All Six Methods

If all the six methods or any two or more of them among *digit exchanging, rotation, rows-in-a-band exchanging, columns-in-a-stack* exchanging, band exchanging, stack exchanging are combined and acted upon on an existing puzzle instance, a new instance can be produced (with no influence on its correctness). This fact is implicit as all these methods in isolation starts from a complete Sudoku solution, and thus, any instance that is created from a complete Sudoku solution must have at least that solution as the outcome of the newly generated instance.

Now here we briefly elucidate a case to show how a new Sudoku instance could be obtained from a given Sudoku instance where several such methods are taking their role in a sequence. Say, first of all, we replace digits 1, 2, 3, 4, 5, 6, 7, 8, and 9 by 7, 3, 6, 5, 4, 8, 2, 9, and 1, respectively, of the Sudoku puzzle shown in Fig. 14b and then rotate the obtained puzzle at an angle of 90° in anticlockwise direction. Then we pair wise exchange rows 4 and 6, rows 7 and 8, columns 1 and 2, and also columns 8 and 9 one after another. In the end, we swap the first band with the second band, and the second stack with the third stack in sequence. After execution

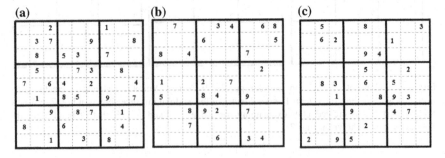

Fig. 14 a A new Sudoku instance is created after exchanging the values in stack 1 and stack 3 present in the Sudoku instance shown in Fig. 7. **b** A given Sudoku instance. **c** A newly generated Sudoku instance after application of several methods of transformation

of all these stages, a new solution grid as an instance of a Sudoku puzzle is created, which is shown in Fig. 14c.

After generating the Sudoku instance it is checked for solutions. We need to check the total number of solutions that exist for the instance. For this we have adapted a graph theoretic technique. In the next section, we are going discuss about the technique.

3.8 A Graph Theoretic Technique to Compute the Number of Solutions for a Produced Sudoku Instance

A Sudoku puzzle can be represented in a simple, symmetrical graph $G = (V, E)$, where the graph structure consists of nine sets of vertices, and each such set comprises and represents a set of valid permutations for a minigrid (of size 3×3 each) of a given Sudoku puzzle P. The graph structure is shown below.

In Fig. 15, each node is represented by a valid permutation of a minigrid. The valid permutation is generated for each and every minigrid and connectivity is given if two valid permutations (of two row and column minigrids) are compatible with each other. The algorithm is as follows.

The Algorithm at a Glance:

Step 1   Compute the number of digits given as clue and the missing digits in each minigrid.

Step 2   Compute all valid permutations of the missing digits for the blank locations (in ascending order) for each minigrid based on the existing clues in P, and store them.

Step 3   Construct a graph $G = (V, E)$, where V is the set of vertices such that a vertex $v_{ix} \in V$ represents a valid permutation p_{ix} of minigrid M_i and $x \geq 1$ is an integer. Thus, altogether for p valid permutations of all nine minigrids in P, $p = |V|$.

Step 4   Two vertices v_{ix} and v_{jy} corresponding to two valid permutations p_{ix} and p_{jy} of a pair of row or column minigrids M_i and M_j are connected by an edge $\{v_{ix}, v_{jy}\} = e \in E$, only if the permutations are compatible to each other, where $1 \leq i, j \leq 3$ and $x, y \geq 1$ are integers.

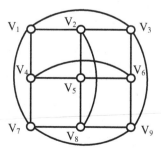

Fig. 15 A Symmetrical graph structure used for representing a Sudoku puzzle

Step 5 Delete vertices with degree three or less in G along with their adjacent edges, and repeat this step until a vertex in G is found with degree less than four; thus, a modified (graph) G is obtained with degree of each vertex four or more.

Step 6 Extract the subgraphs each of which is isomorphic to that as shown in Fig. 15. The number of such distinct subgraphs is the number of solutions for the Sudoku puzzle P.

Using this algorithm we can eventually compute the number of solutions of a created Sudoku instance. Based on total number of solutions of a created Sudoku instance we may categorize the Sudoku puzzles generated having only one solution instance or multiple solution instance that finds applications in different domains of research [11, 12]. In this categorization, we may divide the instances created by the following three: (1) Instances each having a single (or unique) solution, (2) Instances with number of solutions varying from two to four, and (3) Instances with number of solutions of five or more. Still it is a part of future research where computation of the number of solutions for a given Sudoku instance might get immense importance.

4 Conclusion

In this paper, we have developed a new technique that is comprehensive in generating a Sudoku instance based on the difficulty level and the (total) number of solutions possible for a Sudoku puzzle. Our Sudoku instance generator use either the *Digging Hole* strategy or any method of transformation, or a combination of some (or all) of them. From a solved Sudoku puzzle digits are dug in some fashion, either arbitrary or following some desired well-defined sequence. In the method of transformation, we have created a new Sudoku instance by taking as input an existing Sudoku puzzle, and then transformations are applied on it in creating a new one of probably some other level of difficulty. In this paper, we have also introduced a graph theoretic formulation for computing all solutions for a generated instance, which is unique in many a sense.

References

1. Lee, W.-M.: Programming Sudoku. Apress, USA (2006)
2. Jussien, N.: A-Z of Sudoku. ISTE Limited, USA (2007)
3. http://sudopedia.enjoySudoku.com/
4. http://www.sandwalk.blogspot.in/2007/06/i-knew-it-there-can-be-more-than-one.html
5. http://www.mathoverflow.net/questions/27361/do-actual-Sudoku-puzzles-have-a-unique-rational-solution

6. Yato, T., Seta, T.: Complexity and completeness of finding another solution and its application to puzzles. IEICE Trans. Fundam. Electron., Commun., Comput. Sci. **86**(5), 1052–1060 (2003)
7. Herzberg, A.M., Ram Murty, M.: Sudoku squares and chromatic polynomials. Not. AMS **54**, 708–717
8. http://www.math.cornell.edu/∼mec/Summer2009/Mahmood/More.html
9. Stuart, A.C.: Sudoku creation and grading. Mathematica **39**(6), 126–142 (2007)
10. Maji, A.K., Pal, R.K.: An exclusive graph theoretic technique to develop a minigrid based guessed free Sudoku solver. Manuscript (2015)
11. Maji, A.K., Roy, S., Pal, R.K.: A novel steganographic scheme using Sudoku. In: Proceedings of IEEE International Conference on Electrical Information and Communication Technology (ICEICT 2013), pp. 116–121. Bangladesh (2014)
12. Maji, A.K., Pal, R.K.: A novel biometric template encryption scheme using Sudoku. Appl. Comput. Secur. Syst. **2**, 109–128 (2014) (Springer)

Implementing Software Transactional Memory Using STM Haskell

Ammlan Ghosh and Rituparna Chaki

Abstract Software transaction memory (STM) is a promising programming abstract for shared variable concurrency. This paper presents a brief description of one of the recently proposed STM and addresses the need of STM implementation. The paper also describes the implementation technique of STM in STM Haskell. In the STM implementation process, three different approaches have been presented which employ different execution policies. In the evaluation process, transactions with varying execution length are being considered which are executed in multi-threaded environment. The experimental results show an interesting outcome which focuses on the future direction of research for STM implementation.

Keywords Software transactional memory (STM) · Haskell · Concurrency

1 Introduction

Software Transactional memory (STM) [1] is a promising approach for concurrency control in multi-core processors environment. A transaction in STM executes a series of reads and writes to shared memory, which is grouped into an atomic action. STM guarantees that every action will appear to be executed atomically to the rest of the system.

There are several STM approaches those have worked on basic concurrency implementation for avoiding deadlock. These approaches use blocking [2–4] or non-blocking [5–8] process synchronization technique. In non-blocking process synchronization, the major challenge is reducing abort of concurrently executing transactions. A limited of works have been done in this area [7, 8]. In [7], aborting of

A. Ghosh (✉) · R. Chaki
University of Calcutta, Kolkata, India
e-mail: ammlan.ghosh@gmail.com

R. Chaki
e-mail: rchaki@ieee.org

© Springer India 2016
R. Chaki et al. (eds.), *Advanced Computing and Systems for Security*,
Advances in Intelligent Systems and Computing 396,
DOI 10.1007/978-81-322-2653-6_16

transaction has been identified as a major limitation for STM solutions. The work in [7] is on abort- free execution for a cascade of transactions. Although, theoretical estimation shows a good performance improvement; however, the actual STM implementation was not being done to explore the actual performance improvement.

Some STM solutions explore software engineering aspects either by using realistic concurrent data [9, 10, 11] or by a theoretical study [12]. One of the major breakthroughs is the implementation of composable software transactional memory [10, 11] in Haskell.

STM Haskell [10] provides composable memory transaction, i.e., transactional actions that are defined can be combined to generate a new transaction. STM Haskell takes an action as its argument and performs it atomically by maintaining two guarantees: Atomicity and Isolation. Atomicity ensures that execution of a transaction is visible to other threads all at once. Isolation property guarantees that execution of a transaction cannot be affected by other transactions. Since its introduction, several extensions to the basic primitives have been proposed in STM Haskell. This makes STM Haskell more flexible and easy customizable implementation.

This paper describes an implementation of software transactional memory using STM Haskell, using three different concurrency control mechanisms and compares their performance.

The paper is organized as follows: Section 2 presents a brief description of one of the recently proposed STM solution [7] that has claimed to improve throughput in all possible scenarios. This section follows a critical observation on the said work [7] and its analysis to establish the importance of implementing STM solutions on a suitable platform towards appropriate performance analysis. Section 3 describes the implementation technique of software transactional memory using STM Haskell. Section 4 explores the performance and presents the result set. We have presented a set of observations on the advantages of STM Haskell towards implementing STM solutions. The paper concludes in Sect. 5 with a note on future direction of research for STM implementation.

2 Retrospection of an OFTM Solution Towards Abort Freedom

In [7], an interesting obstruction free implementation of STM was proposed that allows contentious transactions to execute without causing any abort to other transactions. The basic idea of this algorithm is that, a transaction, say T_k, may be in active state even after the completion of update process of a transactional variable. Now if another transaction, say T_m, wants to access the same transactional variable, it faces contention with T_k. Thus, in conventional method, either T_m will be blocked or T_m will abort T_k to get access of that transactional variable. In contrast, this algorithm [7] allows T_m to access the transactional variable optimistically, with an expectation that T_k will not update that transactional variable further, thus T_m will find a consistent data value at commit time. At commit time T_m will check the data consistency, i.e., transactional value at the start time is same as at the time of its commit. If data is consistent and T_k is committed, then T_m commits; otherwise, T_m re-executes its operation after reading the last updated value of the transactional variable.

The paper elaborately explains how to execute read and write operations for two transactions in presence of contention. This procedure is also extensible for a cascade of transactions. The efficiency and performance improvement is compared with DSTM (STM for Dynamic-sized Data Structures) [5] in terms of the average execution time (AET) of the transactions. Three different cases are being considered: Where AET of two transactions are equivalent; AET of first transaction less than the second transaction, and lastly, AET of first transaction is greater than the second transaction. The result set shows the throughput of the algorithm is better or at least equivalent to the DSTM [5]. In spite of having several potentials, the algorithm in [7] suffers from some serious drawbacks.

- the solution [7] does not ensure isolation property as transactions communicate between themselves and share the non-committed transactional data;
- the paper [7] proposes abort-free execution, which is tailored only for two concurrent transactions. It has given only an idea on how cascade of transaction may run without any abort;
- the algorithm [7] claims to execute in abort-free manner. However, in some specific cases, transaction either aborts its enemy transaction or backs-off for some arbitrary time;
- authors of [7] claimed that the approach yields higher throughput in comparison to DSTM [5]. However, the actual STM implementation is not done.

These drawbacks can be actually verified and analyzed by implementation or at least by some proper simulation of STM. The GHC STM Haskell could be one of the suitable platforms for STM implementation due to the following reasons:

- GHC Haskell implements some major extensions to support both concurrent and parallel programming, which is highly desirable in multi-core processor environment;
- No new language construct has been introduced in concurrent Haskell, rather it appears as libraries. The functions are exported by these libraries;
- In Haskell, the STM library includes various features like Atomic blocks, Transactional Variables and more importantly the composability of transactions.

All these features make STM Haskell a promising language construct for STM implementation. The algorithm presented in [7] has encouraged the authors of the current paper to discuss on implementing STM in Haskell. In next section, the implementation technique of STM using STM Haskell has been described with different concurrency control mechanism available in GHC Haskell.

3 Implementing STM Using STM Haskell

3.1 Important System Variables

The STM Haskell uses a monad to encapsulate all access operation to shared transactional variables (TVars). The operations in TVars are as follows:

```
data TVar a
newTVar :: a -> STM (TVar a)
readTVar :: TVar a -> STM a
writeTVar :: TVar a -> a -> STM()
```

Both `readTvar` and `writeTVar` operations return STM actions, which can be composed by **do {}** syntax. STM actions are executed by a function atomically, with type `atomically :: STM a -> IO a`.

This function takes memory transaction and delivers I/O action. It runs the transaction atomically with respect to all other transactions. An STM expression can also retry to deal with blocking, when a transaction has to wait for some conditions to be true.

```
retry :: STM a
```

The semantics of retry is to abort the current transaction and run it again. But, instead of blindly rerunning the transaction again and again, transaction reruns only when the TVar that has been read by the current transaction has changed.

Finally, the orElse function allows two transactions to be combined, where only one transaction is performed but not both.

```
orElse :: STM a -> STM a -> STM a
```

The operation `orElse` T_1 T_2 has the following behavior:

- First T_1 is executed, if it returns result then `orElse` function returns.
- If T_1 retry instead then T_1 is discarded and T_2 is executed.

3.2 Implementation

To implement STM in Haskell, we have chosen three different implementation approaches to execute a specific task. The task is to read a sharable data object, calculate the Fibonacci value, and finally write that Fibonacci value to the sharable data object. The whole job will be executed with the protection of atomically.

The first approach uses `TVars` of STM Haskell. The atomically function of STM Haskell maintains a per-thread log that records the tentative access made to `TVars`. Whenever atomically is invoked, it checks whether log is valid, i.e., no concurrent transactions has committed conflicting updates. If the log is valid then transaction commits; otherwise, transaction re-executes with a fresh log.

The next two approaches use TMVars. In Haskell; MVars, Mutable Variables, can be either empty or full. When a value is tired to remove from an empty MVar or to put a value into a full MVar, the operation is being blocked. TVar is modeled with MVar that contains Maybe a, i.e. newtype TMVar a = TMVar (TVar (Maybe a)). 'Maybe a' is very common data type used in Haskell, where a function may or may not succeed. This data type is as follows:

```
Data Maybe a = Nothing
             | Just a
```

The TMVar implementation is included in the Control.Concurrent. STM.TMVar module of STM package in Haskell.

The second approach uses TMVar to execute as per the shortest job first process implementation. The third approach also uses TMVar to execute transactions sequentially in a first-in-first-out basis.

The first approach uses non-blocking synchronization, where as other two use blocking methodology of STM. All these three implementations guarantee atomicity and isolation properties of STM.

Finding Fibonacci Value

Haskell's Control.Parallel module provides a mechanism to allow users to control the granularity of parallelism. The interface is shown below:

```
par :: a -> b -> b
pseq :: a -> b -> b
```

The function par evaluates the first argument in parallel with the second argument by returning its result to the second argument. The function pseq specifies which work of the main thread is to be executed first. The expression a pseq b evaluates a and then returns b. An elaborated explanation on Haskell parallelism is discussed in [13, 14].

While calculating Fibonacci value, the par and pseq monad is used to gain parallelism. The code is as follows:

```
nFib :: Int -> Int
nFib n | n <= 2      = 1
       | otherwise = par n1 (pseq n2 (n1 + n2 ))
              where n1 = nFib (n-1)
                    n2 = nFib (n-2)
```

Achieving Concurrency in Haskell

Haskell provides explicit concurrency features via a collection of library functions. The module Control.Concurrent provides an abstract type ThreadId to identify the Haskell thread. A new thread is created in the IO monad by calling the forkIO function, which returns IO unit.

```
forkIO :: IO () -> IO ThreadId
```

At the time of execution, while using TVars, the main thread in Haskell does not wait for its children threads to complete. The mapConcurrently function has been used to overcome this problem. This function is provided by Haskell's Control.Concurrent.Async module. The function mapConcurrently ensures that main thread does not quit till all its children threads complete their operations. The detailed explanation about this module is available in [14].

STM implementation using TVar

Function transTest is created to define the task of the transaction. The block of code is as follows:

```
transTest :: TInt -> Int -> IO ()
transTest n t = do
  atomically $ do
      let x = nFib t
      writeTVar n x
```

TInt is an integer type Transactional Variable. The type is defined as

```
type TInt = TVar Int
```

The function transTest has two parameters, a TVar and an integer. It calculates Fibonacci value of the given integer and writes that value to the TVar. Calculation of Fibonacci value determines the execution time of the transaction. As nFib 40 takes much more time than nFib 20, thus execution time in the prior case will be higher.

The code for function main (), is as follows:

```
main :: IO ()
main = do
  n <- newTVarIO 0
  _ <- mapConcurrently (transTest n) [40, 20]
```

This code executes two transactions concurrently, where first one will write Fibonacci value of 40 to the TVar *n* and the second one will write Fibonacci value of 20. Now question is that how the Haskell STM will execute these two transactions. As the execution time of the second transaction is less, it completes its execution earlier than first one and finds a valid log value, thus commits. As a result, first transaction will get invalid value in its per-thread log and thus it will re-execute its operation with a fresh log value.

Now suppose we want to track the commit pattern of the transactions. To do so, a list of MVar data type is to be created, where the threadIds will be stored when transactions successfully commit. The modified code is as follows:

```
type TInt = TVar Int

transTest::MVar[(ThreadId, Int)] -> TInt -> Int -> IO ()
transTest mvar n t = do
  tid <- myThreadId
  atomically $ do
      let x = nFib t
      writeTVar n x
  list <- takeMVar mvar
  t2 <- atomically $ readTVar n
    putMVar mvar $ list ++ [(tid, t2)]
main :: IO () -- Asynchronous Thread
main = do
  n <- newTVarIO 0
  ms <- newEmptyMVar
  putMVar ms []

  _ <- mapConcurrently (transTest ms n) [40, 20]
  mms <- takeMVar ms
  putStrLn (show mms)
```

Steps for program compilation

The command to compile the program [13] in multi-threaded environment is as follows:

`$ ghc –o testTVar --make testTVar.hs –threaded –rtsopts`

To execute the program, we need to specify how many real threads are available to execute the logical threads in the Haskell program. The command to execute the program with two real threads is:

`$ ./testTVar +RTS –N2 –s`

The flag –*s*, if included, shows the actual evaluation thread executions. The portion of the actual output, while executing with two threads, is as follows:

```
[(ThreadId 6,6765),(ThreadId 4,102334155)]
   INIT    time    0.00s  (  0.00s elapsed)
   MUT     time    7.25s  (  3.63s elapsed)
   GC      time    1.00s  (  0.50s elapsed)
   EXIT    time    0.00s  (  0.00s elapsed)
   Total   time    8.25s  (  4.13s elapsed)
  Alloc rate   3,611,846,709 bytes per MUT second
  Productivity  87.9% of total user, 175.6% of total elapsed
```

The output shows that commit pattern of the transactions. The execution time is 4.13 s against actual 8.25 s. It also shows the 175.6 % productivity.

STM Implementation using TMVar (Shortest Job First)

In our attempt to implement this, we have used TMVar and threads together. We have created an empty TMVar and forked the job to run in the background. The main thread has been blocked until each results return. While calling Fibonacci, BangPatterns [15] is used to evaluate the Fibonacci value, so that at the time of execution, first thread to finish will have its result first.

We have taken the advantage of TMVar's empty/full semantics to block the main thread for each of the children threads. The program code is given below. The function nFib is same as above. In this implementation also, transactions run atomically and obey the basic principles of STM.

```
{-# LANGUAGE BangPatterns #-}
main :: IO ()
main = do
      result <- newEmptyTMVarIO
      forkIO $ do
        atomically $ do
          let !x = nFib 40
          putTMVar result x
      forkIO $ do
        atomically $ do
          let !x = nFib 20
          putTMVar result x

      t <- atomically $ takeTMVar result
      putStrLn ("Fastest job is: " ++ (show t))

      t <- atomically $ takeTMVar result
      putStrLn ( "Slowest job is: " ++ (show t))
```

STM Implementation using TMVar (First-In-First-Out)

The implementation is same as the previous one, but in this case transactions execute in first-in-first-out basis. Here, the second transaction waits till the first one completes its execution. In this approach, transaction variables are also accessed atomically.

```
main :: IO ()
main = do
        result <- newEmptyTMVarIO
    forkIO $ do
      atomically $ do
        let x = nFib 40
        putTMVar result x

    t <- atomically $ takeTMVar result
        putStrLn ("First value: " ++ (show t))

    forkIO $ do
      atomically $ do
        let x = nFib 20
        putTMVar result x

    t <- atomically $ takeTMVar result
        putStrLn ( "Second value: " ++ (show t))
```

In the next section, these three approaches are being implemented in multi-threaded environment to analyze result set.

4   Simulation Results

In this experiment, parallelism and concurrency both are taken care of while implementing Software Transactional Memory in STM Haskell. This case study considers that transactions perform 'some task', which can be executed in parallel and update the transactional variables. The whole task is to be executed atomically, i.e., either all at once or none. The execution length of transaction depends on the execution time of the task. Thus, throughput of the concurrent execution of transaction also depends on the efficiency of the parallel and concurrent execution of the task.

In this case study three different approaches, as stated in Sect. 3.2, are being considered. The first one is STM Haskell by using TVar, second one (SJF) uses TMVar while execution shortest job first, and third one (FIFO) also uses TMVar but execution pattern is in first-in-first-out basis. The performance of these implementations varies due to these execution policies although all of them ensures STM properties.

The experimental results are summarized by varying execution length of the transaction. To do so, a set of transactions with different execution length are being considered while they are executing concurrently and sharing a common resource. Each set of transaction is formed up with five write transactions. Depending on the AET, transactions are segregated into three groups. In the first group, the AET of transactions is comparatively lower. In the second group, AET is comparatively medium and in third group the AET of transactions is comparatively higher.

In order to investigate scalability, the said three approaches are being executed on these three different groups of transactions. While executing the program, the number of threads is varied from 1 to 5 to observe the efficiency of each method in terms of parallel and concurrent execution.

This implementation is performed on Intel Core i7, 64 bit processor with 8 GB memory, and 2 MB L2 cache, running on Linux and GHC 7.8.3.

4.1 Case-I: Lower Average Execution Time

When transactions have lower execution time, SJF performs best up to three threads. Although, with a higher number of threads, STM Haskell has the slightly better throughput. Table 1 and Fig. 1 show these scenarios.

4.2 Case-II: Medium Average Execution Time

In the case of medium length transactions, performance varies with number of threads, same way as stated in previous case, i.e., with higher number of threads

Table 1 Performance of the said approaches with lower average execution time

	#Thread 1	#Thread 2	#Thread 3	#Thread 4	#Thread 5
STM Haskell	3.92	2.14	1.52	1.31	1.16
TMVar SJF	3.78	2.05	1.45	1.34	1.14
TMVar FIFO	4.01	2.15	1.60	1.42	1.33

Fig. 1 Performance graph of the said approaches with lower average execution time

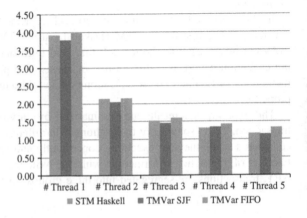

Table 2 Performance of the said approaches with medium average execution time

	#Thread 1	#Thread 2	#Thread 3	#Thread 4	#Thread 5
STM Haskell	8.57	4.66	3.30	2.96	2.62
TMVar SJF	8.36	4.55	3.23	3.19	2.72
TMVar FIFO	9.13	5.00	3.57	3.27	3.12

Fig. 2 Performance graph of the said approaches with medium average execution time

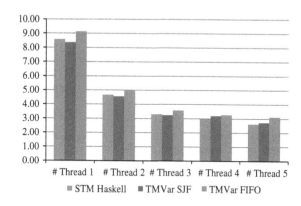

STM Haskell performs better. Table 2 shows the result and Fig. 2 depicts the performance graph.

4.3 Case-III: Higher Average Execution Time

When transactions are too lengthy, STM Haskell outperforms others, except in single-threaded execution. The result set and corresponding graph are shown in Table 3 and Fig. 3 respectively.

4.4 Productivity Improvement with Parallel Execution

Figure 4 shows the average productivity improvement in elapsed time while executing transaction with a varying number of threads. Higher number of threads shows higher productivity.

Table 3 Performance of the said approaches with higher average execution time

	#Thread 1	#Thread 2	#Thread 3	#Thread 4	#Thread 5
STM Haskell	29.46	9.30	11.06	9.69	8.70
TMVar SJF	29.04	15.88	11.13	10.94	9.03
TMVar FIFO	31.35	17.64	13.05	11.19	9.67

Fig. 3 Performance graph of the said approaches with higher average execution time

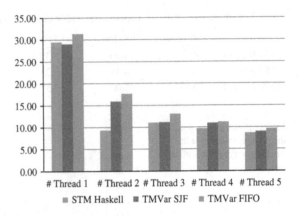

Fig. 4 Productivity improvement on elapse time with increasing number of threads

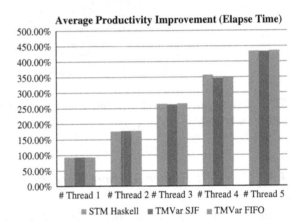

4.5 Summary of Results

The shortest job execution policy has minimum waiting time, which implies low turnaround time for processes. For this reason, in single-threaded environment, our implementation performs better with shortest job first execution policy. In multi-threaded environment, the parallel activities, i.e., scheduling the job for multi-cores, switching between threads etc. are managed by Haskell compiler and OS. Under this scenario, our STM implementation with TVar performs better than other two approaches. When transactions' execution length is higher, this approach performs best while executing in multi-threaded environment. In our third implementation, where transactions execute in first-in-first-out basis, transactions' average waiting time becomes higher, which results in high turnaround time and a lower throughput.

5 Conclusions

In this paper, we have critically described one of the recently proposed STM solutions to establish the importance of STM implementation for appropriate performance analysis. We have also implemented STM in Haskell using three different approaches. The first implementation uses TVars (STM Haskell) to access transactional variables concurrently. Each transaction maintains a log and depending on its validity transaction re-executes with a fresh log. In second implementation, we have combined TMVar and bang-pattern for strict evaluation, which enables transactions to execute any job in the background. Using this technique, we implemented shortest job first execution policy. The third implementation executes transactions as per their initiation order in first-in-first-out basis.

In all these implementations, we have executed task of the transactions in parallel and observed the performance impact of different execution policies. The experimental results show variations in performance depending on number of threads and transactions' execution length. Transactions with smaller execution length perform better in shortest job first implementation when number of threads is less. When number of threads is increased, the STM Haskell performs better. On the other hand, when transaction execution length is high, STM Haskell performs better, irrespective of number of threads available.

References

1. Shavit, N., Touitou, D.: Software transactional memory. In: ACM SIGACT-SIGOPS Symposium on Principles of Distributed Computing, pp. 204–213. ACM (1995)
2. Harris, T., Fraser, K.: Language support for lightweight transactions. In: OOPSLA '03: Proceedings of the 18th Annual ACM SIGPLAN Conference on Object-Oriented Programing, Systems, Languages, and Applications, pp. 388–402. ACM Press (2003)
3. Gidenstam, A., Papatriantafilou, M.: LFthreads: a lock-free thread library. In: Principles of Distributed System, pp. 217–231. Springer, Berlin (2007)
4. Fernandes, S.M., Cachopo, J.: Lock-free and scalable multi-version software transactional memory. In: ACM SIGPLAN Notices, vol. 46, no. 8, pp. 179–188. ACM (2011)
5. Herlihy, M.P., Luchangco, V., Moir, M., Scherer, W.M.: Software transactional memory for dynamic-sized data structures. In: Proceedings of the 22nd Annual Symposium on Principles of Distributed Computing (2003)
6. Tabba, F., Wang, C., Goodman, J.R., Moir, M.: NZTM: non-blocking zero-indirection transactional memory. In: Proceedings of the 21st ACM Annual Symposium on Parallelism in Algorithms and Architectures (SPAA), pp. 204–213. ACS (2009)
7. Ghosh, A., Chaki, N.: The new OFTM algorithm toward abort-free execution. In: Proceedings of the 9th International Conference on Distributed Computing and Information Technology, pp. 255–266. Springer, Berlin (2013)
8. Dolev, S., Fatourou, P., Kosmas, E.: Abort Free SemanticTM by Dependency Aware Scheduling of Transactional Instructions, Preprint, TRANSACT'13 (2013)
9. Discolo, A., Harris, T., Marlow, S., Jones, S.P., Singh, S.: Lock free data structures using STM in Haskell. In: Functional and Logic Programming, pp. 65–80. Springer, Berlin (2006)

10. Harris, T., Marlow, S., Peyton Jones, S., Herlihy, M.: Composable memory transactions. In: PPoPP 2005. ACM Press, New York (2005)
11. Du Bois, A.R.: An implementation of composable memory transactions in Haskell. In: Software Composition, pp. 34–50. Springer, Berlin (2011)
12. Borgström, J., Bhargavan, K., Gordon, A.D.: A compositional theory for STM Haskell. In: Proceedings of the 2nd ACM SIGPLAN Symposium on Haskell, pp. 69–80. ACM (2009)
13. Jones, S.P., Singh, S.: A tutorial on parallel and concurrent programming in Haskell. In: Advanced Functional Programming, pp. 267–305. Springer, Berlin (2009)
14. Marlow, S.: Parallel and Concurrent Programming in Haskell, 1st, edn. O'Reilly Media, Inc. (2013)
15. O'Sullivan, B., Goerzen, J., Stewart, D.: Real World Haskell, 1st edn. O'Reilly Media, Inc. (2008)

Comparative Analysis of Genetic Algorithm and Classical Algorithms in Fractional Programming

Debasish Roy, Surjya Sikha Das and Swarup Ghosh

Abstract This paper compares the performances of genetic algorithm with various classical algorithms in solving fractional programming. Genetic algorithm is one of the new forms of algorithms for solving optimization problems, which may not be efficient but a generic way to solve nonlinear optimization problems. The traditional optimization algorithms have difficulty in computing the derivatives and second order partial derivatives, i.e., Hessian for the fractional function. The issues of discontinuity seriously affects traditional algorithm. There are large numbers of classical methods for searching the optimum point of nonlinear functions. The classical search algorithms may be largely classified as gradient based methods and nongradient methods. Here, a comparative performance analysis of different algorithms is made through a newly defined function called algorithmic index. An algorithm based on heuristics for computation of gewicht vector required to derive algorithmic index has also been proposed here.

Keyword Fractional programming · Genetic algorithm · Optimization · DEA

1 Introduction

Rechenberg, a German scientist, introduced evolutionary strategies for airfoil shape point optimization, in the 1960s. Fogel, Owen, and Walsh formulated evolutionary programming for finite state machines. Evolutionary computation is a broad field

D. Roy (✉) · S.S. Das
Department of Management Studies, Techno India University, Kolkata, India
e-mail: debasishroy7@gmail.com

S.S. Das
e-mail: surjyasikha.tiu@gmail.com

S. Ghosh
Department of Humanities and Social Sciences, Techno India University, Kolkata, India
e-mail: swarupghosh55.tiu@gmail.com

© Springer India 2016
R. Chaki et al. (eds.), *Advanced Computing and Systems for Security*,
Advances in Intelligent Systems and Computing 396,
DOI 10.1007/978-81-322-2653-6_17

and remains a serious interest for research with sub areas as evolutionary programming, evolutionary strategies, and genetic algorithm. Large number of researchers worked on evolution-based algorithms, notable among them are Box (1957); Friedman (1959); Bledsoe (1961); Bremermann (1962); Reed, Tombs, and Baricelli (1967).

Genetic algorithm, developed by John Holland who published a book titled 'Adaptation in Natural and Artificial Systems' in 1960 in the University of Michigan, mimic adaptation and traverses from one set of population to a new set of population by crossover, mutation and selection to achieve and find fitter, and more suitable population. Later schemas formed basis for all subsequent developments. This paper compares performances of traditional methods like Random Search, Box Evaluation, Gradient Descent method, and Hookes' Jeeves method with Genetic Algorithm on the Fractional Programming. In order to have an even comparison, same fractional function is taken for experimentation and optimized values have been derived using Matlab scripts. This paper also proposes algorithmic index for comparison of performances of these algorithms. The computation of algorithmic index depends on gewicht vector; estimation of this vector can be done by newly proposed heuristics dependent algorithm.

2 Literature Review

The optimization of the ratio of linear functions has attracted researchers for many years. In many practical applications, multiple such fractions need to be optimized. Von Neumann was the earliest to have started using fractional programming in Equilibrium problems, in 1937. The linear fractional programming started with B. Mortars and his associates. Charnes and Cooper [1] had proposed transformation LFP to Linear Programming format and thereby solving the problem. Bitranes and Novaes solved the problem by gradient descent method [2]. Bitranes and Magnant [3] made analysis in the same score as Linear Programming for duality and sensitivity. Swarup [4] extended simplex method to LFP, which is an extension of work of Danjtzig [5, 6], for solving LFP. Bounded variable linear fractional programming was explored by Bazalinov [7]. He also worked on scaling problems in LFP [8]. Interval Valued Fractional Programming was studied by Shohrab Effati [9]. Fractional Programming has often been addressed in the domain of generalized fractional programming [10].

Until 1980s, the single ratio problem has dominated the field. Various methods have evolved; important among them are use of the LPP methods for solving fractional functions. Gogia [11] suggested revised simplex method for solving fractional programming. Horvath [12] gave a criteria for determining optimum for linear fractional programming using duality considerations. A finite method for solution to a simpler form of the fractional programming was developed by Stahl [13]. A special class of linear fractional programming is fractional interval programming, which has been studied by a number of authors [14]. Buhler solved fractional interval programming using generalized inverses [15].

2.1 Random Search Method

Random search method uses a population created either in random manner or by performing unidirectional search along the random search direction. If the feasible space is narrow, the system of random search may fail. Random Search method has various algorithms starting from Pure Random Search and Random Walk to Metaheuristic. Pure Random Search [16] is a stochastic search method of selecting random population based on normal or Gaussian distribution. Convergence can be proved in number of ways [17–19]. The procedure may guarantee convergence, but is not efficient as it may require large number of iterations depending on nature of the objective function, feasible zone, and dimension of variables. It may be enormously large. One such Random Search algorithm is presented in supplementary material.

2.2 Box's Evolutionary Method

Introduced by G.E.P Box in 1957, this evolutionary algorithm, centering on a point, selects best point among 2 N points evaluated along N dimensions. The next iteration starts from this point. The size of the hypercube is successively reduced with change of the position of the center. Here, required number of function evaluation increases exponentially with N. This is one of the strong drawbacks of the Box's Method. There are other versions of Box's Evolutionary Method like Random Evolutionary Operation (REVOP), Simplex EVOP, etc. The convergence of the algorithm depends on initial hypercube size, position, and reduction rate, ∂_i. This is basically a multi-dimensional systemic search on dimensions [20]. The algorithm is useful since it is a derivative free optimization method (DFO) [21–23]. Subsequently, further improvement was made by Wilson [24] and it was renamed as Response Surface Methodology (RSM). Whereas RSM is based on least square, DFO is based on direct function values or its interpolation. The hypercube is the focal point for searching optimum point. The Box's method was also opted in finding optimum solutions in industrial applications [25]. The algorithm, however, does not guarantee convergence to local or global optima. The algorithm is presented in supplementary material.

2.3 Hooke Jeeves' Method

In 1961, Hooke and Jeeves [26] conceived that direct search method is effective when the objective function is nondifferentiable or does not have derivative at all points in feasible region. In this method, each trial is compared with the previous best [27]. Therefore, direct search methods for unconstrained optimization works on relative rank of countable set function values whereas, Armijo-Goldstein-Wolfe condition for quasi-Newton line search algorithm requires a sufficient decrease in objective function. The algorithm is presented in supplementary material.

2.4   Gradient Descent Method

Here, one of the two popular gradient descent methods is presented. The direct search method eventually requires large number of steps or iterations to converge, whereas gradient-based methods are faster. However, convex optimization methods define subdifferentials for functions having discontinuity, which is defined as follows:

$$\partial f\left(\bar{\bar{x}}\right) = \left\{v \in R^n : f(y) - f\left(\bar{\bar{x}}\right) \geq \left\langle v, y - \bar{\bar{x}}\right\rangle, \forall y \in R^n\right\}. \tag{1}$$

In cases where the objective function is differentiable, the gradient methods [28] or derivative-based methods are useful and efficient, compared to direct search method. One of the oldest systems of finding multivariable optimum points is Newton's Method [29]. This method is extremely important as it is the simplest method and assures convergence [30]. Large numbers of improved algorithms have emerged from this algorithm with slight or minimum modification. One of the most accepted methods based on the Newton Method and that has been widely accepted is the Conjugate Direction Method [31]. A generalized powerful extension of this is the Spacer Step Theorem [32]. The algorithm for Gradient Descent Method is given in supplementary material.

2.5   Genetic Algorithm

The schema based on genetic algorithm was developed in 1968, with famous disposition of Schema Theorem [33]. Goldberg [34] developed building block hypothesis from Schema Theorem. The criticism of Schema Theorem developed in 1990 that the effect of noise and other stochastic effects distort proportionate selection [35, 36, 37].

The algorithmic flow is given in the supplementary material. The parameters of genetic algorithm are

- Cross Over Probability—It is 0 %, if the offspring population is an exact copy of the parent. It is 100 %, if all of the parent population is allowed to crossover.
- Mutation Probability—Mutation probability is zero, if no population is changed after cross over. It is 100 %, if the whole chromosome is changed. Mutation is necessary to prevent falling of the population in local optimum.
- Population Size: If population size is small, the search space will not be covered well. If search space is large, the algorithm becomes slow. Population size also determines the precision of the solution, i.e., the quality of the solution.

Besides choosing appropriate population size, the balance between selection operator and exploration operator introduced by crossover and mutation operator is also important. If selection operator uses too much selection pressure, then the population loses diversity. Genetic algorithm is used for NP-hard problems. It is more robust than conventional algorithm, i.e., the algorithm does not collapse in the presence of noise or change of inputs. Also, genetic algorithm is useful in searching n-dimensional surface or multimodal search space.

3 Linear Fractional Programming

Linear Fractional Programming has various forms. Let p, q, and s denote real valued functions which are defined over C ∈ R^n. Let us take

$$s(x) = \frac{p(x)}{q(x)}. \tag{2}$$

The function s is defined over $D = \{x \in C : s(x) \leq 1\}$ assuming $q(x) \neq 0$ for x ∈ C.

Single Ratio Fractional Programming may be defined as

$$\text{Max}\{s(x) : x \in D\}. \tag{3}$$

In many practical applications, multiple ratios appear for evaluation. This is also referred to as max−min problem. The Generalized Fractional Programming may be defined as

$$\text{Max}\{\min s_i(x) : x \in D\}, \text{ where } s_i(x) = \frac{p_i}{q_i}, \text{where } i = 1, 2 \ldots m \text{ and } s_i > 0. \tag{4}$$

We call it as concave fractional programming, if numerator p_i is concave on D. The denominator q_i is convex function on D. It is further assumed that p_i is non-negative on D, if q_i is not affine. The objective function in general is not assumed to be concave. The objective function is assumed to be a ratio of convex and concave function. The fractional programs are, in general, assumed to be nonconcave programs. The central point of fractional programming is objective function and point of attraction is the ratio structure with a feasible region being a convex polyhedron.

Sometimes, functions in both numerator and denominator are affine functions. If D is a convex polyhedron, the problem is called Linear Fractional Program. The form of the function is as follows:

$$Max\left\{\frac{a^T x + \emptyset}{b^T x + \theta} : Ax \leq c, x \geq 0\right\}, \text{Where } a, b \in R^n, \emptyset, \theta \ R, A \in R^{mxn}, c \in R^m. \tag{5}$$

4 The Experiment

In this experiment, a standard two variable (0–1) fractional function is taken as follows:

$$\text{Maximize} f(x, y) = \frac{3x + 4y + 1}{5x + 7y + 5} \left\{ \forall (x, y) \in \mathcal{R}^2 : 5x + 7y + 5 \neq 0 \right\}. \qquad (6)$$

During the experiment, the optimum value that is the maximum value of this sample fractional function is obtained. The goal is to maximize the fraction. This is treated as unconstrained optimization problem. There are large numbers of methodologies available for finding the optimum value of this NP-hard problem. Here, three nonderivative-based methods and one derivative-based method is chosen. Genetic algorithm is also used to find the optimum value. The comparison of the computation time, number of iterations, and optimum values of all the algorithms is found. The parameter sensitivity and other studies can also be conducted. These algorithms are primarily numerical methods; as a result, comparison of efficiency of algorithms analytically is difficult. The efficiency of the algorithms depend not only on parameters but also on type of function. The experiment has been conducted largely on same setting, which is the same function and same computer. The precision of output is also chosen same in all cases. The basic nature of the algorithms prevents complete identical context generation. As a result, some intrinsic differentiation remains.

4.1 Random Search

Here, 1000 random values are chosen between 0 and 1, and function value is evaluated at all these points. The maximum of the function value is chosen. The random numbers are chosen based on normal distribution. The random numbers are 2-dimensional. The Matlab code is given below (Fig. 1)

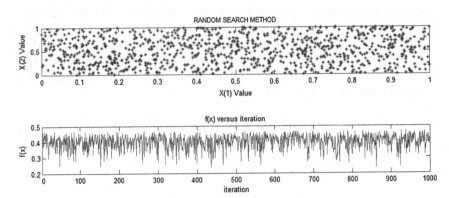

Fig. 1 The plots of iteration points and f(x,y) versus iteration

```
function
[foptxopttimespent]=rando
mOptim(iter)
tic;
xran=ones(iter,2);
f=ones(1,iter);
fori=1:iter
xran(i,:)=rand(1,2);
    X=xran(i,:);
f(i)=fx1(X);
holdon;
subplot(2,1,1);
The plot of output is:
```

```
plot(X(1),X(2),'*');
holdoff;
title('RANDOM SEARCH
METHOD');
end
[fopt p]=max(f);
xopt=xran(p,:);
subplot(2,1,2);
plot(f);
timespent=toc;
end
```

The output is as follows:

$$\text{fopt} = 0.4698, \text{xopt} = 0.99950.9810, \text{timespent} = 30.7039$$

4.2 *Box Evolutionary Method*

In the case of Box's Evolutionary method, initial point is chosen as (x,y) = (2,3). The precision or delta value is chosen as 0.01. The Matlab code is given below (Fig. 2)

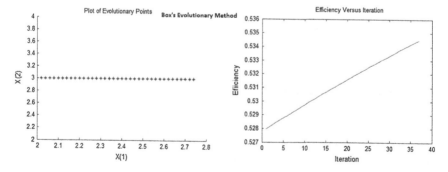

Fig. 2 The plot of (x,y) and f(x,y) with iterations

```
function                          end
[fmaxXmaxiteriterTime]=bo         subplot(1,2,2);
xEval(xinit,d)                    plot(E(1:iter));
clc;                              Xmax=y;
tic;                              fmax=Emax;
iter=0;                           iterTime=toc;
x=xinit;                          end
delta=5;                          function [EmaxEmin  y]=
Xset=ones(100,2);                 maxbox2(x,d)
E=ones(100);                      Val=Eff2(x);
while(delta > .0001)              if Val < 1.0
iter=iter+1;                      x1=[x(1)+d x(2)];
    [EmaxEmin                         E1=Eff2(x1);
y]=maxbox2(x,d);                  x2=[x(1)+2*d x(2)];
delta=Emax-Emin;                      E2=Eff2(x2);
    x=y;                          x3=[x(1) x(2)+d];
Xset(iter,:)=y;                       E3=Eff2(x3);
E(iter)=Emax;                     x4=[x(1) x(2)+2*d];
end                                   E4=Eff2(x4);
disp(Xset);                           E=[E1;E2;E3;E4];
plot(Xset);                           [Emax,I]=max(E);
fori=1:iter                       Emin=min(E);
subplot(1,2,1);
holdon;                           y=eval(strcat('x',num2str
plot(Xset(i,1),Xset(i,2),         (I)));
'*');                             end
holdoff;                          end
```

The output is as follows:

fmax = 0.5345, Xmax = 2.74003.0000, iter = 37, iterTime = 0.2862

4.3 Hooke's Jeeves Pattern Search

In this optimization method, the combination of exploratory and heuristic move is used to find the optimum value. The initial starting point is chosen as (x,y) = (2,3). The delta and alpha values are chosen as 0.1 and 2. The Matlab code is given below (Fig. 3)

```
function
[funcmaxXmaxitertimespent
]=hj(X,delta,alpha)
tic;
clc;
cleardata;
delta1=delta;
f=ones(100);
fnext=0;
iter=0;
fmax=.1;
funcmax=0;
while (abs(funcmax-
fmax)>.0001)
iter=iter+1;
funcmax=fmax;
    [fmaxXmax
t]=expl(X,delta1);
if t==1
        X1=X;
X2=Xmax;[fnextXnext]=patt
ern(X1,X2,alpha);
else
delta1=delta/2;
end
X=Xmax;
holdon;
subplot(2,1,1);
plot(X(1),X(2),'*');
title('Hookes Jeeves Plot
of X(1), X(2)');
holdoff;
f(iter)=funcmax;
timespent=toc;
end
subplot(2,1,2);
plot(f(1:iter));
end
function
[fnextXnext]=pattern(X1,X
2,alpha)
test=1;

count=0;
while (test==1)
count=count+1;
Xnext=X2+alpha*(X2-X1);
if ((fx1(X2)>fx1(Xnext))
|| count==2)
Xnext=X2;
test=0;
else test=1;
end
X2=Xnext;
end
fnext=fx1(Xnext);
end
function [fmaxXmax t] =
expl(X, StepSize)
D=StepSize;
X1=[X(1)+D X(2)];
X2=[X(1) X(2)+D];
X3=[X(1)-D X(2)];
X4=[X(1) X(2)-D];
if (X3(1)<0)
    X3(1)=0;
end
if (X4(1) < 0)
    X4=0;
end
f=fx1(X);
f1=fx1(X1);
f2=fx1(X2);
f3=fx1(X3);
f4=fx1(X4);
[fmax I]=max([f f1 f2 f3
f4]);
if I==1
    t=0;Xmax=X;
else
t=1;Xmax=eval(strcat('X',
num2str(I-1)));
end
end
```

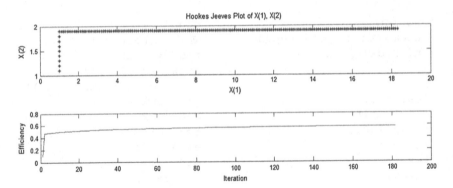

Fig. 3 The plot of (x,y) and f(x,y)

The output is as follows:

$\text{funcmax} = 0.5772, \text{Xmax} = 17.70003.0000, \text{iter} = 157, \text{timespent} = 1.4124.$

4.4 Gradient Ascent Method (Cauchy's Method)

Here, the search direction is the direction of gradient of the function at the point of evaluation in contrast to the negative of gradient in Gradient Descent Method. The Matlab Code is given below

```
function
[fgradmaxxoptitergradtime
]=gradDescent(xinit)
clc;
tic;
iter=0;
[dfx]=gradf(xinit);
f=ones(100);
xiter=ones(100,2);
[alpha]=gS(xinit,dfx);
x1=xinit-alpha*dfx;
Nfx=fx1(x1);
Ofx=fx1(xinit);
if (Nfx>Ofx)
xinit=x1;
end
while((Nfx-Ofx) > .00001)
iter=iter+1;
    [dfx]=gradf(xinit);
    [al-
pha]=gS(xinit,dfx);
x1=xinit-alpha*dfx;
Nfx=fx1(x1);
Ofx=fx1(xinit);
if (Nfx>Ofx)
xinit=x1;
end
f(iter)=Nfx;
xiter(iter,:)=x1;
disp(Nfx);
disp(Ofx);
end
xopt=x1;
fgradmax=Nfx;
gradtime=toc;
subplot(1,2,1);
plot(f(1:iter));
subplot(1,2,2);
plot(xiter((1:iter/100),1
),xiter((1:iter/100),2),'
*');
```

```
end
function [dfx]=gradf(x)
d=.001;
x1=[x(1)+d x(2)];
x2=[x(1)-d x(2)];
x3=[x(1) x(2)+d];
x4=[x(1) x(2)-d];
dfx(1)=(fx1(x2)-
fx1(x1))/(2*d);
dfx(2)=(fx1(x3)-
fx1(x4))/(2+d);
dfx=[dfx(1) dfx(2)];
end
function y=fx1(X)
y=(3*X(1)+4*X(2)+1)/(5*X(
1)+7*X(2)+5);
end
function [al-
pha]=gS(x,dfx)
a1=1;
a0=0;
ah=a1/2;
x1=x+a1*dfx;
xh=x+ah*dfx;
x0=x+a0*dfx;
while ((x1-x0)>.001)
if(fx1(x1)<fx1(xh))
        a1=ah;
end
if(fx1(x0)<fx1(xh))
        a0=ah;
end
ah=a1/2;
x1=x+a1*dfx;
xh=x+ah*dfx;
x0=x+a0*dfx;
end
alpha=a1;
end
```

The plots of number of iterations and the function value with respect to the iterations are given below (Fig. 4).

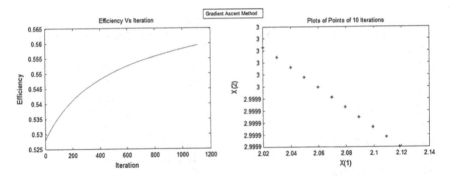

Fig. 4 The plots of (x,y) and f(x,y) against iterations

The output is as follows:

gradmax = 0.5595, Xopt = [7.63152.9965], iterations = 1113, gradtime = 0.3190

4.5 Genetic Algorithm

In case of genetic algorithm, a sample population in feasible region is selected at random. The crossover and mutation within the sample population is performed to get a new population based on the fitness level of sample population. The fitness function is the criteria for survival in subsequent generation. Here, the value of the objective function or Q(x) is the criteria for fitness. The sample population size depends on the accuracy of result expected, i.e., number of bits required for encoding the variable. We assume two decimal place accuracy of the variable. With the increase in number of variables, the cross over and mutation operations become complicated. The termination criterion has been chosen as difference of two successive iterations less than 10^{-4}. The population size has been chosen as 10. The Matlab code is given below (Fig. 5)

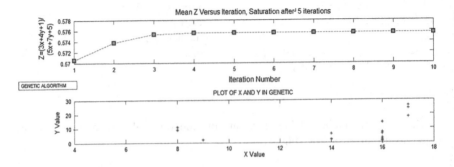

Fig. 5 Plot of output of genetic algorithm

```
function
[ZmaxiterXmaxYmaxtimespen
t]=callgeneFrac(x,y,tol,m
ax_iter)
clc;
tic;
r=randi([8,10]);
k=5;
output=ones(2,10,200);
[x1,y1,z,e,scat]=geneFrac3(x,y,r
,5);
out=ones(1,10);
mut=ones(1,10);
zmean=mean(z);
emean=mean(e);
p=0;
while(abs(zmean-emean)>tol)
    r=randi([9,10]);
    p=p+1;
if (p>max_iter/2)
        k=k+1;

[x1,y1,z,e,scat]=geneFrac3(x,y,r
,k);
end
if(p>2*max_iter/3)
        k=k+1;
[x1,y1,z,e,scat]=geneFrac3(x,y,r
,k);
end
if (p>max_iter)
break;
end
    x=x1;
    y=y1;

[x1,y1,z,e,scat]=geneFrac3(x,y,r
,5);
output(:,:,p)=[x1;y1];
zmean=mean(z);emean=mean(e);
out(p)=zmean;
    [Zmax m]=max(z);
Xmax=x1(m);
Ymax=y1(m);
end
for k=p+1:10
out(k)=zmean;
end
iter=p;
timespent=toc;
figure;
subplot(2,1,1);
plot(out,'--
rs','LineWidth',2,'MarkerEdgeCol
or','k','MarkerFaceColor','g','M
arkerSize',10);
xlabel('Iteration
Number','FontSize',12);
ylabel('Z=(3x+4y+1)/(5x+7y+5)','
FontSize',12);
title(strcat('Mean Z Versus It-
eration, Saturation
after',blanks(4),
iter),'FontSize',12);
for k=1:p
for l=1:10
holdon;
subplot(2,1,2);
plot(output(1,1,k),output(2,1,k)
,'*');
title('PLOT OF X AND Y IN
GENETIC');
holdoff;
end
end
end
```

The output is as follows:

Zmax = 0.5765, iter = 4, Xmax = 16, Ymax = 0, timespent = 0.5549

5 Result

This paper is intended to compare five algorithms. The five algorithms were run with the following Matlab code:

```
tol=.001;max_iter=100;
clc;
disp('Genetic
Algorithm');
[ZmaxiterXmaxYmaxtimespen
t]=callgeneFrac(x,y,tol,m
ax_iter);
tgene=timespent;
itergene=iter;
disp('Hookes Jeeves');
[funcmaxXmaxitertimespent
]=hj([1 1],.1,2);
thj=timespent;
iterhj=iter;
disp('Random Search');
[foptxopttimespent]=rando
mOptim(1000);
trs=timespent;
maxpts=1000;
disp(' Box Evaluation Method');
[fmaxXmaxiteriterTime]=bo
xEval([1 1],0.01);
tbox=iterTime;

iterbox=iter;
disp('Gradient Ascent
Method');
[fgradmaxgraditergradtime
]=gradDescent([1 1]);
fprintf('Computation Time
Genetic=%f, Hookes=%f,
Random Search=%f,
BoxEval=%f Grad=%f\n',
tgene,thj,trs,tbox,gradti
me);
fprintf(' Optimum values Genetic=%f,
Hookes=%f, Random Search=%f, BoxEval=%f
Grad=%f\n',
Zmax,funcmax,fopt,fmax,fg
radmax);
fprintf('Iterations Ge-
netic=%d, Hookes=%d, Ran-
dom
Search=%d,BoxEval=%d,Grad
=%d\n',itergene,iterhj,ma
xpts,iterbox,graditer);
```

The comparative result of running five algorithms individually on the same fractional function with almost same termination criteria and same initial starting point is given below in tabular form (Table 1 and Fig. 6).

Table 1 Comparative figures for various algorithms

Algorithms	Computation time	Iterations	Optimum values	Computation time/iteration
Random search	0.013811	1000	0.47016	0.138113
Box's evolution	0.154882	231	0.536978	6.704833
Hooke's Jeeves	0.14433	191	0.5779115	7.7556552
Gradient ascent	0.186542	1213	0.562720	1.537856
Genetic	1.4901413	52	0.55645161	286.565718

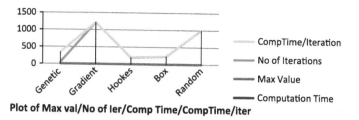

1500
1000
500
0

Genetic   Gradient   Hookes   Box   Random

········ CompTime/Iteration
──── No of Iterations
──── Max Value
──── Computation Time

Plot of Max val/No of Ier/Comp Time/CompTime/iter

Fig. 6 Plot of (computation time/iteration)/optimum value/time per iteration for different types of algorithms

6 Conclusion

From the table of comparative performances, it is clear that computation time for genetic algorithm is highest, while random search takes lowest time. Strangely, Hooke's Jeeves Pattern Search takes lesser time to execute than Box's Evolution and others except Random Search. There is no doubt that the performance of genetic algorithm is the best. Though Gradient Descent is known to be a better algorithm than non gradient algorithms, Hooke's Jeeves algorithm performs better than Gradient Search in terms of computation time. In regard to number of iterations to reach termination, again genetic algorithm performs better than all other algorithms. The worst is the Random Search Algorithm. The performance of Hooke's Jeeves Method is better than Box's Evolutionary algorithm. This compelled the derivation of computation time per iteration. From the computation time per iteration, it is found that Gradient Descent Algorithm is having the best result, whereas genetic algorithm is the worst. This may be due to the fact that more complex is the algorithm, the computation time per iteration is higher.

The most interesting feature in Hooke's Jeeves algorithm is that it is giving the highest value in comparison to other algorithms. Gradient Descent Algorithm is giving the next highest. The performance of Random Search in terms optimum value is worst and highest in case of Hooke's Jeeves Algorithm.

The fractional programming is NP-hard problem. As a result, computation of algorithmic complexity is difficult. This paper does not intend to find the complexity. However, in order to make comparative assessment, Algorithmic Index is defined as follows:

$$\eta = \Delta_1(\text{CT}) + \Delta_2(\text{It}) + \Delta_3(\text{Opt}) + \Delta_4(\text{CT/It}).$$ (7)

CT   Computation Time
It   Iterations
Opt   Optimum Values
CT/It   Computation Time per Iteration
$\Delta = (\Delta_1 \Delta_2 \Delta_3 \Delta_4)$ Relative gewicht vector

Computation of Gewicht Vector: In order to estimate the gewicht vector, the following algorithm has been formulated:

Step 1 Choose random set of initial starting points.
Step 2 Find optimum parameters.
Step 3 Set a heuristics for derivation of Algorithmic Index.
Step 4 Performance Matrix is computed as follows:
        Performance Matrix = [CT_{Algo} It_{Algo} Opt_{Algo} $(CT/It)_{Algo}$],
        , where Algo = Gentic Algorithm, Gradient Ascent Algorithm,
        Hooke's Jeeves, Box, and Random.
Step5   Normalize Performance Matrix.
Step6   Derive Algorithmic Index as follows:
        AlgoIndex = GewichtVector * Normalized Performance Matrix
Step7   Find the set of gewicht for which the heuristics satisfies.
Step8   The average gewicht over the set is computed.
Step9   If the average gewicht does not satisfy heuristics, readjust heuristics
        by going to step2, and recompute till gewicht satisfies heuristics.
Step10 Estimated gewicht vector is derived.
        Step11—Apply estimated gewicht vector to the Normalized
        Performance Matrix on a new random initial starting point. Check
        whether heuristics is satisfied.

The MATLAB code for estimation of gewicht vector is given in supplementary material. The estimated gewicht turns out to be $(\Delta_1 \Delta_2 \Delta_3 \Delta_4) = (0.237586\, 0.292759\, 0.358276\, 0.111379)$

With the random starting point figures, the performance matrix turns out to be Table 2

After normalizing the performance index and using estimated gewicht vector, algorithmic index turns to be:

Table 3

In a nutshell, it may be concluded that genetic algorithm performs far better than other algorithms. The algorithmic index is highest for genetic algorithm. The ratio of η

Table 2 Performance matrix

Performance matrix	Genetic	Gradient	Hookes	Box	Random
Computation time	1.49014173	0.186542	0.14433	0.154882	0.013811
Max value	0.55645161	0.56272	0.577915	0.536978	0.47016
No of iterations	52	1213	191	231	1000
CompTime/iteration	286.565718	1.537856	7.556552	6.704833	0.138113

Table 3 Normalised performance matrix

Normalized performance matrix					
	Genetic	Gradient	Hookes	Box	Random
Computation time	107.892569	13.50643	10.45011	11.21409	1
Max value	0.96286003	0.973707	1	0.929164	0.813544
No of iterations	1	23.32692	3.673077	4.442308	19.23077
CompTime/iteration	2074.85709	11.13473	54.71263	48.54583	1
Algorithmic index	257.37	13.092	10.185	9.935	7.477

values of genetic and gradient, the nearest competitor is approximately 19.658. That means, performance of genetic algorithm is nearly 20 times better than the remaining. The Gradient Ascent performs better than Hooke's Jeeves and Box's Algorithm. Here, two variable fractional functions have been studied. The research may be extended to cases of higher dimensions.

Acknowledgements Special thanks go to my guides Dr. Sujyasikha Das and Dr. Swarup Prasad Ghosh for inspiring me to write the paper and implementing the scenario in Matlab. The paper would remain unfinished if I don't convey my regards and heartfelt thanks to Dr. Nabendu Chaki for relentless support to my academics. He has been the driving force for all the activities.

Appendix

Algorithm for Random Search

Step 1: Choose initial x^0, z^0, ϵ such that the minimum lies in $(x^0 - 1/2z^0, x^0 + 1/2z^0)$. For each Q block, set $q = 1$ and $p = 1$.

Step 2: For $i = 1,2...N$, create points using uniform distribution of m in the range $(-0.5, 0.5)$. Set $x_i^{(p)} = x_i^{q-1} + mz_i^{q-1}$.

Step 3: If $x^{(p)}$ is infeasible and $p < P$, repeat Step 2. If $x^{(p)}$ is feasible, save $x^{(p)}$ and $f(x^{(p)})$. Increment p and repeat step 2;

Else if $p = P$, set x^q to be the point that has lowest $f(x^{(p)})$ over all feasible $x^{(p)}$ including x^{q-1}

And reset $p = 1$.

Step 4: Reduce the range via $z_i^q = \epsilon z_i^{q-1}$.

Step 5: If $q > Q$. Stop.

Else increment q and continue to Step 2.

Box's Evolutionary Algorithm

Step 1: Choose initial point. Choose size reduction step δ_i and termination criteria ϵ.

Step 2: If $\delta_i < \epsilon$ STOP.

Step 3: Else create 2^N points by adding and subtracting δ_i from each variable at the initial point.

Step 4: Compute function values at all 2^N points. Find the optimum among these points. Set it as initial point for next iteration.

Step 5: Reduce size of the step to $\delta_i/2$ and go to Step 2.

Hooke's Jeeves' Algorithm

Step 1: Initial point is selected and objective function is evaluated.

Step 2: Search is made in the direction of each dimension by a step size S_i to find lowest of functional value.

Step 3: In case the function value does not decrease in any direction, the step size is reduced and fresh search is made.

Step 4: If the value of objective function reduces, a new initial point is found as follows:
$$X_{i,o}^{(k+1)} = X_i^{k+1} + \theta(X_i^{k+1} - X_i^k), \, \theta > 1.$$

Step 5: This search continues till the termination criteria is met, i.e., $\theta < \epsilon$.

Gradient Descent Method

Step 1: Choose initial point $x^{(0)}$ and termination parameters ϵ_1 and ϵ_2.

Step 2: Compute first derivative $\nabla f(x^k)$.

Step 3: If $\|\nabla f(x^k)\| \leq \epsilon_1$ STOP.
        Else go to next step

Step 4: By unidirectional search, find α^k such that $f(x^{(k+1)}) = f(x^k - \alpha^k \nabla f(x^k))$ is minimum. One criteria for termination is $|\nabla f(x^{k+1}).\nabla f(x^k)| \leq \epsilon_2$.

Step 5: If $\frac{\|x^{k+1} - x^k\|}{\|x^k\|} \leq \epsilon_1$, then STOP.
        Else set $k = k + 1$, go to step 2.

Genetic Algorithm

Start and generate a random population of size n.

Fitness: Evaluate fitness of each chromosome.

New Population: Create new population by repeating the steps below

Select two parent chromosomes from the population according to best fitness.

Cross over the parents, with a crossover probability to form new population.

With a mutation probability, mutate the new offspring.

Add the new offspring in the population.

Replace: Use the new generation for next iteration.

Test: Check termination criteria.

Loop: Go to step 2.

MATLAB Code for Estimating Gewicht Vector

```
ExtFiveMethods.m
function []=ExtFiveMethods()
  % for computing Parameters
 x=rand(1,10);
 y=rand(1,10);
    [avS avAlgoIndex]=pmComp(x,y);
    x=rand(1,10);
    y=rand(1,10);
    fn='fiveMethod.mat';
    [NPM1 PM]=FiveMethods(x,y);
    save(fn,'NPM1','PM');
    AlgoIndex=avS*NPM1;
    fn1='fiveMethod1.mat';
    save(fn1,'avS','AlgoIndex');
    disp('Estimated Algorithimic Index');
    fprintf('%1.3f %1.3f %1.3f %1.3f %1.3f\n',AlgoIndex);
    fprintf('\n');
end
pmComp.m
function [avS avAlgoIndex]=pmComp(x,y)
x1=x;
y1=y;
[NPM perfMatrix]=FiveMethods(x1,y1);
paramSt=ones(50,4);
t=1;
    for k=0.01:.1:1
        for j=0.01:.1:1
            for i=0.01:.1:1
                l=1-i-j-k;
                param=[i j k l];
                if ((l<=0)||(k==1)||(j==1)||(i==1)||(i==1))
                    break;
                end
                 AlgoIndex=param*NPM;
                if ((i<1) || (j<1) || (k<1) || (l<1))
                        if( (AlgoIndex(1,1)>AlgoIndex(1,2)) &&
...(AlgoIndex(1,2)>AlgoIndex(1,3)) &&
(AlgoIndex(1,3)>AlgoIndex(1,4))...
                    && (AlgoIndex(1,4)>AlgoIndex(1,5)) )
                      paramSt(t,:)=param;
                            t=t+1;
                        end
                end
            end
```

```
                    end
                end
        end
        comps=sum(paramSt(1:t-1,:));
        avS=(1/(t-1))*comps;
        avAlgoIndex=avS*NPM;
        pause;
        end
FiveMethods.m
function [NPM perfMatrix]=FiveMethods(x,y)
tol=.001;max_iter=100;
[Zmax iter Xmax Ymax timespent]=callgeneFrac(x,y,tol,max_iter);
tgene=timespent;
itergene=iter;
 [funcmax Xmax iter timespent]=hj([x(1) y(1)],.1,2);
thj=timespent;
iterhj=iter;
[fopt xopt timespent]=randomOptim(1000);
trs=timespent;
maxpts=1000;
[fmax Xmax iter iterTime]=boxEval([x(1) y(1)],0.01);
tbox=iterTime;
iterbox=iter;
[fgradmax xopt graditer gradtime]=gradDescent([1 1]);
format;
perfMatrix=ones(4,5);
perfMatrix(1,:)=[tgene gradtime thj tbox   trs ];
perfMatrix(2,:)=[Zmax fgradmax funcmax fmax   fopt  ];
perfMatrix(3,:)=[itergene graditer iterhj iterbox  maxpts ];
perfMatrix(4,:)=10000*[tgene/itergene  gradtime/graditer
thj/iterhj tbox/iterbox trs/maxpts ];
tmin=min(perfMatrix(1,:));
OutMax=max(perfMatrix(2,:));
iterMin=min(perfMatrix(3,:));
perIterMin=min(perfMatrix(4,:));
NPM=ones(4,5);
NPM(1,:)=(1/tmin)*perfMatrix(1,:);
NPM(2,:)=(1/OutMax)*perfMatrix(2,:);
NPM(3,:)=(1/iterMin)*perfMatrix(3,:);
NPM(4,:)=(1/perIterMin)*perfMatrix(4,:);
End
```

References

1. Charnes, A.C.W.: An explicit general solution in linear fractional programming. Naval Res. Logist. Quart. **20**, 449–467 (1973)
2. Bitran, G.R., Novaes, A.G.: Linear programming with fractional objective function. Oper. Res. **21**, 22–29 (1973)
3. Birtan, G.R., Magnanti, T.L.: Duality and sensitivity analysis of fractional objective function. Oper. Res. **24**, 675–699 (1976)
4. Swarup, K.: Linear fractional programming. Oper. Res. **13**(6), 1029–1036 (1965)
5. Dantzig, G.B.: Linear Programming under uncertainty. Manage. Sci. **1**(3 and 4), 197–206 (1955)
6. Dantzig, G.B., Mandansky, A.: on solution of two stage linear programming under uncertainty. Barkley Symp. Maths Stat. **1**(3 and 4), 165–176 (1961)
7. Bazalinov, E.B.: Linear Fractional Programming. Kluwer Academic Publishers, Dordrecht (2003)
8. Bajalinov, E.: Scaling problems in linear fractional programming. In: proceeding of 10th international conference on operation research, vol. 3, no. 1, pp. 22–24 (2004)
9. Shohrab, E., Morteza, P.: Solving the Interval Valued Fractional Programming. Am. J. Comput. Math. **2**(1), 51–55 (2012)
10. Barros, A.I., Frenk, J.B., Schaible, S., Zhang, S.: A new algorithm for generalised fractional programming. Math. Program. **72**(2), 147–175 (1996)
11. Gogia, N.: Revised simplex algorithm for linear fractional programming problem. Math. Student **36**(1), 55–57 (1969)
12. Horvath, I.: AsupraprogramliriifracJionare lineare cu restricJii suplimentare. Informatica pentru Conducere, pp. 101–102 (1981)
13. Stahl, J.: Two new methods for solution of hyperbolic programming. In: Publications of the Mathematical Institute of Hungarian Science, vol. 9, no. B, pp. 743–754 (1964)
14. Stancu-Minasian, I.M.: Stochastic Programming with MultiObjective Function. D. Reidel Publishing Company, Dordrecht (1984)
15. Buhler, W.: A note on fractional interval programming. Oper. Res. A-B **19**, 1, 29–36 (1975); **Z**(19), 29–35 (1975)
16. Robins, H., Monro, S.: A stochastic approximation method. Ann. Math. Stat. **22**, 400–407 (1951)
17. Costa, A., Jones, O., Kroese, D.: Convergence properties of the cross entropy method for discrete optimization. Oper. Res. Lett. **35**, 573–580 (2007)
18. Zhang, Q., Muhlenbein, H.: On the convergence of a class of estimation of distribution algorithm. IEEE Trans. Evol. Comput. **8**, 127–134 (2004)
19. Binglsley, P.: Convergence of probability measures. John Wiley and Sons, New York (1999)
20. Box, G.E.: Evolutionary operation: a method for increasing industrial productivity. Appl Stat **6**, 81–101 (1957)
21. Brent, R.P.: Algorithms for Minimization without derivatives. Printice Hall, EngleWoods Cliffs (2002)
22. Mifflin, R., Strodiot, J.J.: A Bracketing technique to ensure desirable convergence in univariate minimisation. Math. Prog. **17**, 100–117 (1975)
23. Mifflin, R., Strodiot, J.J.: A rapidly convergent five-point algorithm for univariate minimisation. Math. Prog. **62**, 299–319 (1993)
24. Box, G.P., Wilson, K.B.: On the experimental attainment of optimal conditions. Stat. Soc. **13**, 1–13 (1951)
25. Box, G.E.P., Draper, N.R.: Evolutionary Operation: A Statistical Method For Process Improvement. Wiley, New York (1998)
26. Hooke, R., Jeeves, T.A.: Direct search solution for numerical and statistical problem. ACM **8**, 212–219 (1961)

27. Nelder, J.A., Mead, R.: A simplex method for function minimisation. Comput. J. **7**, 308–313 (1965)
28. Fletcher, R.: Practical Methods for Optimisation. John Wiley and Sons, Chichester (1987)
29. Murray, W., Wright, M.H., Gill, P.E.: Practical Optimization. Academic Press, London (1981)
30. Gabay, D.: Reduced Quasi Newton method with feasibilty improvement for nonlinear constrained optimisation. Math. Prog. Stud. **16**, 18–44 (1982)
31. Fletcher, R.: Conjugate Gradient Methods for Indefinite Systems. Numer. Anal. Rep. **5**, 11 (1975)
32. Zangwill, W.: Nonlinear Programming: A Unified Approach. Printice Hall, Englewood Cliffs (1969)
33. Holland, J.H.: Hierarchical description of universal spaces and adaptive systems. Tech. Rep ORA Project 01252 (1968)
34. Goldberg, D.: Genetic Algorithms in Search, Optimization, and Machine Learning. Addison-Wesley, Reading (1989)
35. Grefenstette, J.J.: Deception considered harmful. In: Whitley, L.D. (ed.) Foundations of genetic algorithms (1993)
36. Fogel, D.G.: Schema processing under proportional selection in the presence of random effects. IEEE Trans. Evol. Comput. **1**(4), 290–293 (1997)
37. Radcliffe, N.J.: Schema Processing. In: Handbook of evolutionary computation, pp. B2.5–1.10. Oxford University Press (1997)

An Algorithm to Solve 3D Guard Zone Computation Problem

Ranjan Mehera, Piyali Datta, Arpan Chakraborty
and Rajat Kumar Pal

Abstract The guard zone computation problem finds vast applications in the field of VLSI physical design automation and design of embedded systems, where one of the major purposes is to find an optimized way to place a set of 2D blocks on a chip floor. Each (group of) circuit component(s) C_i is associated with a parameter δ_i, such that a minimum clearance zone of width δ_i is to be maintained around C_i. In this paper, we introduce the problem in its 3D version. Considering 3D simple solid objects makes the guard zone computation problem more complex and helps to solve many real life problems like VLSI physical design, Geographical Information System, motion control in robotics, and embedded systems. In this paper, we develop an algorithm to compute guard zone of a 3D solid object detecting and excluding overlapped regions among the guard zonal regions, if any.

Keywords Simple polygon · Safety zone · Notch · Convex hull · False hull edge · Convolution · Minkowski sum · Extreme points of a curve · 3D simple solid object · Computational geometry · 3D coordinate geometry

R. Mehera (✉) · P. Datta · A. Chakraborty · R.K. Pal
Department of Computer Science and Engineering, University of Calcutta, Acharya Prafulla Chandra Roy Siksha Prangan, JD – 2, Sector – III, Saltlake City, Kolkata 700098, West Bengal, India
e-mail: ranjan.mehera@gmail.com

P. Datta
e-mail: piyalidatta150888@gmail.com

A. Chakraborty
e-mail: arpanc250506@gmail.com

R.K. Pal
e-mail: pal.rajatk@gmail.com

© Springer India 2016
R. Chaki et al. (eds.), *Advanced Computing and Systems for Security*,
Advances in Intelligent Systems and Computing 396,
DOI 10.1007/978-81-322-2653-6_18

271

1 Introduction

Guard zone computation problem is well defined in literature as an application of Computational geometry. Often, this problem is known as safety zone problem [1]. In case of 2D guard zone computation problem, given a simple polygon P, its guard zone G (of width r) is a closed region consisting of straight line segments and circular arcs (of radius r) bounding the polygon P such that there exists no pair of points p (on the boundary of P) and q (on the boundary of G) having their Euclidean distance $d(p,q)$ less than r. In case of VLSI layout design as well as in embedded system, a chip may contain several million transistors. The goal of placement is to find a minimum area arrangement for the blocks that helps to complete interconnections among them. A good routing and circuit performance heavily depend on a good placement algorithm. Placement of modules is an NP-complete problem.

Each circuit component P_i is associated with a parameter p such that a minimum clearance zone of width p must be maintained around that circuit component. The location of the safety zone of specified width for a simple polygon is an important problem for resizing the circuit components. If more than one polygonal region is close enough, their safety zones overlap, violating the minimum separation constraint among them. Thus, with respect to resizing problems in VLSI, this is the motivation of defining the safety zone of a polygon [2].

We have developed a number of algorithms to solve the guard zone computation problem for only 2D simple polygons or objects. Now, the question arises whether the problem can be visualized and solved for its 3D version. A 3D simple solid object is surrounded by a number of planes such that no two planes cut each other except at their edges. The pair of planes meeting at an edge is the neighboring planes. In this paper, we develop an algorithm to compute guard zone of a 3D solid object detecting and excluding overlapped regions among the computed guard zonal regions, if any.

If two or more objects are close enough so that their guard zones overlap, indicating the violation of the minimum separation constraint among them, the intersecting regions are to be detected such that the guard zone can be computed eliminating those intersecting regions. As our inclination in doing the task is in the domain of computational geometry for a given 3D simple object, we like to detect the part(s) of G that overlap(s) using the concept of analytical and coordinate geometry.

In this paper, we have solved the problem of computation of the guard zone for a simple solid object as well as detection of the overlapped regions (if any). While computing the initial guard zone G, we enclose the solid object P by G, which is essentially a collection of O(n) planar segments, cylindrical segments, and spherical components corresponding to the planar surfaces, convex edges, and convex solid vertices of the input simple solid object; we explain it in the subsequent sections. After computing the guard zone trivially, we detect the overlapped regions of the guard zone to find the union of all individual guard zonal regions of the object. Again, the 3D guard zone computation algorithm proposed in this paper is output sensitive in nature, i.e., the computational complexity varies with the number of overlapped regions in the guard zone of the solid object. Hence, the complexity depends on the shape of the simple solid object provided.

2 Literature Survey

If P is a simple polygon and G is its guard zone of width r, then the boundary of G is composed of straight line segments and circular arcs of radius r, where each straight line segment is parallel to an edge of the polygon at a distance r apart from that edge, and each circular arc of radius r is centered at a (convex) vertex of the polygon. The boundary of the guard zone describes a simple region in the sense that no two edges (straight line segment(s) and/or circular arc(s)) on its boundary intersect in (or pass through) their interior. The problem originates in the context of resizing the VLSI layout design [3].

In the context of guard zone computation, several different algorithms have been proposed so far. The most discussed tool for guard zone computation is the Minkowski sum. Essentially, Minkowski sum between a line (as polygonal segment) and a point (perpendicularly at a distance r apart) with same x- and y-coordinates gives a line parallel to the given one. But a question arises whether the parallel line is inside or outside the polygon. Here, the definition of Minkowski sum [4] can be extended as follow: if A and B are subsets of R^n, and $\lambda \in R$, then $A + B = \{x + y \mid x \in A, y \in B\}, A–B = \{x–y \mid x \in A, y \in B\}$, and $\lambda A \{\lambda x \mid x \in A\}$. Note that $A + A$ does not equal to $2A$, and $A–A$ does not equal to 'zero' in any sense. Apart from Minkowski sum, convolution can also be used as a tool for guard zone computation.

A linear time algorithm is developed for finding the boundary of the minimum area guard zone of an arbitrarily shaped simple polygon in [3]. This method uses the idea of Chazelle's linear time triangulation algorithm [5]. After having the triangulation step, this algorithm uses only dynamic linear and binary tree data structures.

Again, the problem of locating guard zone of a simple polygon has been solved and a time-optimal sequential algorithm for computing a boundary of guard zone that uses simple analytical and coordinate geometric concepts have been presented in [6, 7]. It uses three different procedures to compute guard zone at convex, concave, and linear regions of the polygon. The algorithm can easily be modified to compute the regions of width r outside the polygon (as guard zone), and also inside the polygon.

In paper [8], the authors have developed algorithm for detection of guard zonal overlapping in case of a 2D simple polygon and the algorithm uses the concept of line sweep algorithm for a set of parallel line segments. Most interestingly, the algorithm is output sensitive, i.e., its behavior changes with the input. Thus, the guard zone computation is easier for a polygon without notches than that of a polygon with notches as well as overlapped guard zonal regions.

3 Formulation of the Problem and the Algorithm

In case of 2D guard zone computation problem [8], our objective is to derive a 2D imaginary region outside the polygon such that each point p on the polygon maintains at least a distance r, which is predefined from each point q on that region. Similarly, for the 3D version of the guard zone computation problem, we derive a 3D imaginary region bounding the solid object such that it maintains at least distance r between them. For the sake of simplicity, we consider only that kind of solid objects, which consist of only planes, i.e., no curved surfaces are there. Hence, two neighboring planes meet at their edges, i.e., at a straight line where the corresponding planes make an angle that may either be convex or concave as it is formed outside the object. If the planes meet at a convex angle outside the object, such an edge is considered to be a convex edge; otherwise, it is considered as a concave edge if the planes meet at a concave angle outside the object.

On the other hand, a simple solid object may contain both convex and concave vertices in it; at such a vertex, we call a solid vertex, several planes of the solid object intersect. It is important to observe that at each such vertex of a simple solid object, the number of planes intersects is at least three. The vertices of a simple solid object are defined as follows. A (solid) vertex v of a solid object S is defined as concave, if for each pair of intersectional lines (for the associated planes) incident at v form an angle outside the object (i.e., an external angle at vertex v) less than 180°; otherwise, it is defined as convex.

As in Fig. 1a, which is a portion of solid object S where two adjacent planes A and B intersect, the normal vector for plane B is n_1 and that of A is n_2. Now, the angle between planes A and B is same as the angle between the normal vectors to planes A and B. Hence, $\theta = \cos^{-1}(|n_1 \cdot n_2|/(|n_1||n_2|))$. This is how all external angles of the solid object S can be computed (as concave or convex) in time $O(\ell)$, where ℓ is the number of intersection lines between adjacent planes of S. Now, it can be said that, if at a solid vertex the number of convex edges meeting at the vertex is greater than the number of concave edges, the vertex is a convex solid vertex; whereas, it is

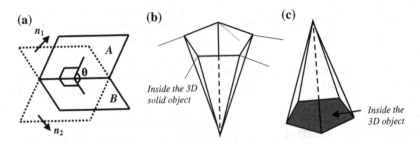

(a) **(b)** **(c)**

Inside the 3D solid object

Inside the 3D object

Fig. 1 **a** Part of a solid object with planes A and B, and a concave (external) angle θ between them. The *dotted lines* indicate the imaginary portion of the planes A and B (inside the solid object); n_1 and n_2 indicate the normal vector to the planes B and A, respectively. **b** Deep or concave solid vertex in a 3D solid object. **c** Peak or convex solid vertex in a 3D solid object

said to be a concave solid vertex if the number of concave edges meeting at the vertex is greater than that of convex edges meeting at the vertex. A concave and a convex solid vertex have been depicted in Fig. 1b, c, respectively.

A guard zone G of a 3D simple solid object S with n solid vertices can eventually be obtained as follows. Let intersection lines created due to the intersection of adjacent planes of the object be labeled as $l_1, l_2, ..., l_\ell$ in some order. For each intersection line l_i, $1 \leq i \leq \ell$, where two adjacent planes of S intersect, we bisect the external angle. Then we draw a plane parallel to the plane (l_i, l_{i+1}), $1 \leq i < \ell$, at a distance r outside the solid object that may be a portion of the desired guard zone G that is being computed, assuming that l_i and l_{i+1} are forming the plane under consideration. To be precise, successive guard zonal planes must meet (or intersect) each other only at an angle bisector of line l_i, formed due to the intersection of allied planes of the given 3D simple solid object. At the edges of the simple object, the guard zonal region is cylindrical in shape. Furthermore, the parallel planes that are guard zone of the neighboring planes are tangent to the cylindrical surface whose axis is the line of intersection of the two assumed planes. At the peak, where more than two planes coincide, the guard zonal regions result a spherical shape and the guard zonal planes of the given object's planes that meet at the peak are tangent to the spherical surface.

As the input 3D simple solid object is made of a set of planes (by assumption), for computing the guard zone of individual plane, we first need to compute planes parallel to the ones specified in the form of 3D simple solid object, at a distance r outside the object. For example, we have considered two planes for which we would like to compute parallel planes at distance r outside the two believed planes. Let $ABCD$ and $BEFC$ are two planes adjacent through the edge BC as shown in Fig. 2a. Since we always consider a simple solid object as input, in reality these two planes are also adjacent to some other planes of the input solid object through different edges.

For $ABCD$, we draw four perpendicular line segments at A, B, C, and D of length r. Thus, we get Aa, Bb, Cc, and Dd, respectively, and obtain the plane $abcd$

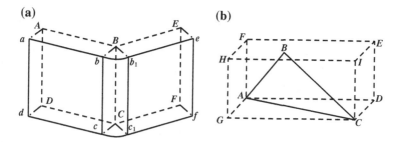

Fig. 2 **a** Two plane segments ($ABCD$ and $BEFC$) meet at a line segment (BC) and their guard zones ($abcd$ and b_1efc_1) meet at a cylindrical segment (bb_1c_1c). **b** Planar surface ABC bounded by the 3D box $AGCDFHIE$

which is parallel to the plane *ABCD* at distance *r*. Similarly, for the plane *BEFC* we get plane b_1efc_1 as its parallel one.

Now, the guard zonal planes of two neighboring object planes meet at a cylindrical surface which is considered to be the intermediate curved surface between the two planes said above. To compute this surface, we have drawn a cylindrical surface considering *BC* as its axis and making an angle, $\angle bBb1$ at the axis. Now, the guard zonal planes of the object planes, i.e., abcd and $b1efc1$, are tangent to this cylindrical surface.

Now, there is a set of guard zonal components in the search space from which we have to find out the pair of intersecting components. In case of 2D, i.e., the sample space contains only a set of line segments, we could use plane sweep algorithm to find out the interesting pairs of line segments. However, in 3D, space sweep algorithm is only applied for a set of orthogonal planes though the guard zonal plane segments are not necessarily orthogonal [9]. Furthermore, the search space also contains cylindrical and spherical regions. Hence, space sweep algorithm cannot be applied directly.

Let us take a different view toward the problem. If we could bind each guard zonal component within its minimum possible orthogonal 3D box, then the problem reduces to find the overlapping pairs of those boxes only. Again, a 3D orthogonal box consists of six bounding surfaces parallel to one of the three coordinate planes and the problem is reduced in finding overlaps among these boxes which are regular in shape.

Now, we can imagine that in the search space there are only $O(n)$ 3D orthogonal boxes, where n is the total number of planes in a given 3D simple solid object. Exhaustively, $O(n^2)$ checking needs to be performed to find all the intersections among the set of boxes. Instead of the exhaustive method just described, we like to use the space sweep method [9], which solves the problem in $O(n \log n)$ time. This three-phase sequential algorithm computes $O(n)$ number of 3D bounded boxes in its first phase and in the second phase it checks for overlapping among the boxes, whereas in the third phase it deals only with the corresponding guard zonal regions for which overlapping has been detected in the second phase. Thus, if there is no overlapping between any two boxes, the third phase of the algorithm is skipped and the results we obtain are reported accordingly.

Phase I: Construction of Bounded Boxes for Individual Guard Zonal Surfaces
As the guard zone of a 3D simple solid object may consist of a set of planar, cylindrical, and spherical surfaces, we compute bounded boxes for each such surface individually. To bind a planar or a curved surface, we take orthogonal projection of each surface on xy, yz, and zx planes of the coordinate system. Hence, in each plane we obtain either a line segment or a curved segment or a bounded region which are not necessarily identical. Thus, we get three 2D counterparts of each surface of the 3D guard zone. For each of the 2D segments, we compute its 2D bounded box as described in [8]. Now, we obtain three mutually perpendicular planes, i.e., three surfaces of the 3D bounded box. The three other surfaces also need to be constructed to complete the 3D bounded box of the guard zonal portion

taken into consideration. We like to illustrate this concept with the help of an example discussed below.

Let us consider a guard zonal planar surface ABC as shown in Fig. 2b, of which we would like to construct the bounded box. At first, we take orthogonal projection of ABC on xz, xy, and yz planes. This creates three 2D regions on the three planes for which we find three 2D bounded rectangles. Here, $AGHF$, $ADCG$, and $ADEF$ are three bounded rectangles on the three coordinate axis planes, respectively. Hence, the three adjacent planes are computed of the 3D box and other three planes are yet to be constructed. The width of the 2D box on yz plane indicates the width of the 3D box along y axis and z axis; similarly, we get the width along x axis from the 3D box formed on planes xz and xy.

Now, to construct the remaining surfaces of the 3D box, we have to draw three planes parallel to xz, xy, and yz planes. For an example, the plane $DCIE$ is drawn parallel to the plane $AGBF$ maintaining a distance AD which is the width of the 2D box on yz (or xy) plane as well as the width of the 3D box along y axis. Similarly, the face $GCIH$ and $HIEF$ of the 3D box are drawn parallel to $ADEF$ and $ADCG$, respectively. Thus, we get the 3D box $AGCDEFHI$, which bounds the guard zonal plane ABC. We perform similar task to get 3D bounded boxes for each of the guard zonal (planar or curved) surfaces. The 3D bounded boxes computed in this fashion are not necessarily disjoint to each other, i.e., they may have overlaps. Figure 3a, b shows the projection of a spherical surface on three orthogonal coordinate planes.

Phase II: Detection of Overlapping among the Bounded Boxes

At the end of the first phase, there are $O(\ell)$ number of 3D boxes in the search space, in which there may be overlap among the boxes, where ℓ denotes the number of intersection lines at which the consecutive plane segments of the object meet. In the second phase, we use the space sweep algorithm [4] that checks for intersection among the boxes and report it accordingly.

In the search space, an infinite plane parallel to each of the xz, xy, and yz planes is moved along its perpendicular direction (i.e., along y-, z-, and x-axis) consecutively. For each sweep, we get information regarding overlapping of the boxes along the corresponding direction. Suppose two boxes overlap while sweeping

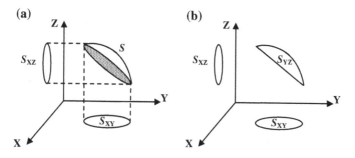

Fig. 3 **a** Projections are taken on the orthogonal coordinate planes **X**. **b** Projection of the spherical surface S on xy plane (S_{XY}), yz plane (S_{YZ}), and xz plane (S_{XZ})

through y axis; however, it does not necessitate having overlapping along other two axes. It means that there was no overlapping between them; rather, they share their y-interval. By Lemma 1, we can conclude that two boxes overlap if and only if overlapping has been detected along all the three directions.

Lemma 1 Two boxes overlap if and only if they share x-, y- and z-interval.

Proof Let B1 and B2 be any two boxes in the search space. The x-, y-, and z-span of B_1 and B_2 are $\{(x_{11} - x_{12}), (y_{11} - y_{12}), (z_{11} - z_{12})\}$ and $\{(x_{21} - x_{22}), (y_{21} - y_{22}), (z_{21} - z_{22})\}$, respectively. In Fig. 4, we observe that $x_{11} < x_{21}$, $y_{11} < y_{21}$, and $z_{11} < z_{21}$. Now in a case, if any two boxes overlap, then the point(s) belonging to the overlapped region is (are) also belonging to the individual boxes, and hence, if we that consider boxes B_1 and B_2 have overlapped and an arbitrary point A(x, y, z) is a point within the overlapped region, then the following inequalities hold.

$$x_{i1} < x < x_{i2}, \ y_{i1} < y < y_{i1}, \text{ and } z_{i1} < z < z_{i1}, \text{ where } i = 1, 2. \quad (1)$$

Considering the fact stated in the inequation (1), we can conclude that $x_{21} < x_{12}$, $y_{21} < y_{12}$, and $z_{21} < z_{12}$, i.e., there are overlaps along all x-, y-, and z-direction. On the flip side, if there is no overlapping between boxes B_1 and B_2, the inequation (1) stated above would not be satisfied and must deny the existence of any such point A, which in turn ensures that there is no overlapping between the mentioned boxes.

Now, in our algorithm for overlapping detection among the 3D bounded boxes, we store the information of overlapping for three different directions and boxes by maintaining three different Binary Search Tree (BST) data structures. Again, we extract the final information regarding overlapping by combining the results of these three BSTs.

We illustrate the process through an example. Let us consider that there are three 3D boxes, namely 1, 2, and 3 on which we perform space sweep operations along three coordinate axes. The surfaces of the boxes which are parallel to the sweep plane are considered to be the event points during sweeping. For an example, when we move the xz plane, the surfaces of the boxes parallel to xz plane are considered to be the event points and the surface with lower y value is the starting point of the allied box while the surface with higher y value is considered to be the end point. At the beginning of the sweeping process, the event points are sorted depending on the values of that coordinate along which the sweeping is being performed. Through

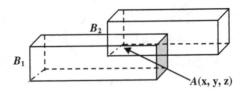

Fig. 4 Two boxes B_1 and B_2 are overlapped and A(x, y, z) is a point at the overlapped region

the sweeping process whenever the sweep plane is at a starting point, the corresponding box is inserted in the query tree and the overlapping list is updated by inserting overlapping information of the newly inserted box and the existing boxes in the query tree. On the other hand, at an end point the corresponding box is deleted from the tree. Overlapping information for two boxes is stored in the form of a dipole {box_1, box_2}.

Let us consider the example depicted in Fig. 5a–c. In Fig. 5a, xz plane is moving along y axis and the event points have been denoted as 1s, 1e, 2s, 2e, 3s, and 3e. Now, the event points are sorted according to their y-coordinate values and we obtain 2s, 3s, 1s, 2e, 1e, and 3e. At each event point, the query tree is updated through insertion or deletion and the event point is deleted from the event point list. Hence, the sweeping ends when the event point list is empty and we obtain a list of overlapping 3D boxes.

In Fig. 5a, when the sweep plane is at the starting plane 2s, the corresponding box_2 is inserted in query tree. Next event point is 3s and it is also a starting point; hence, box_3 is inserted as the right child of box_2, as its y-coordinate value is greater than that of box_2. The overlapping list which was initially empty is now updated by inserting the pair of boxes {2, 3} or {3, 2}. To remove ambiguity, we prefer to store the lower numbered box first, i.e., {2, 3} is inserted into the overlapping list. The next event point is 1s and box_1 is inserted into the query tree as the right child of box_3.

After the insertion, as the tree becomes height imbalanced, AVL rotation is performed to balance the tree. Now the overlapping list is updated by inserting event points {1, 3} and {1, 2}. At the next three event points, as these are the end points, the allied boxes are deleted from the query tree. Finally, the query tree as well as the event point list becomes empty and we obtain a set of overlapping pairs along y axis, say A. Here, A = {{1, 2}, {2, 3}, {1, 3}}. The query tree after each update has been shown in Fig. 6a.

Next, the sweep plane is moved through z axis and the surfaces of the boxes that are parallel to xy plane are considered to be the event points. Now, the starting and ending event points are shown in Fig. 5b; after sorting we obtain the sequence of the event points as 3s, 2s, 3e, 1s, 2e, and 1e and the query trees are updated at each event point. After completion of the sweeping, the set of overlapping pairs of boxes, say set B, becomes {{2, 3}, {1, 2}}. The series of query trees formed after each operation performed for each event point has been shown in Fig. 6b. At the last step of the space sweep phase, yz plane is swept along x axis where the sorted list contains 1s, 2s, 1e, 2e, 3s, and 3e, the event points based on Fig. 5c.

At the end of sweeping, it provides the overlapping information through a set of overlapping pairs. If the set is named as C, C = {{1, 2}}. The series of query trees formed after each operation performed for each event point has been shown in Fig. 6c.

We have already proved the phenomenon that if two 3D boxes have overlapped with each other, they must overlap along all the three axes. Now, to detect the boxes which overlap indeed, we have to find the common pair(s) in these three sets A, B, and C. In our example, if we perform A∩B∩C, we obtain only one pair {1, 2},

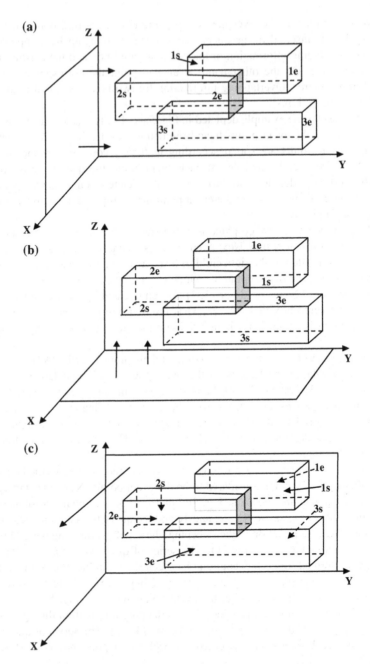

Fig. 5 Sweep of the plane along **a** y axis, **b** z axis, and **c** x axis

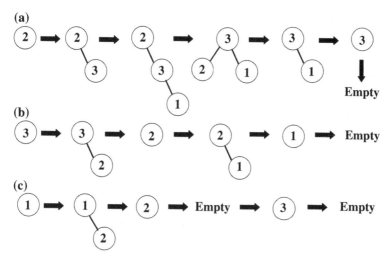

Fig. 6 Query trees during the sweep of the plane along **a** y axis, **b** z axis, and **c** x axis

though there are three entries in set A and two entries in set B. Hence, box1 and box2 have intersection and we deal with only these boxes in the third phase, as only these boxes contain the probable intersecting guard zonal surfaces.

Phase III: Detection of Intersection among the Guard Zonal Components Contained in the Overlapped Boxes

The third phase of our algorithm deals with those guard zonal surfaces whose bounded boxes are found overlapped in the second phase. As there are only three types of guard zonal surfaces, any pair of them may intersect with the other, and there are only six types of possible intersections, that are planar–planar, planar-spherical, planar-cylindrical, spherical-spherical, spherical-cylindrical, and cylindrical-cylindrical.

Planar–Planar Intersection If we like to check intersection between two plane segments, we notice that two plane segments always cut at a line segment satisfying the equation of both the planes [10]. Two plane segments intersect in two ways; either one of them fully passes through the other or they intersect partially. If A and B are two plane segments, the possible intersections are depicted in Fig. 7.

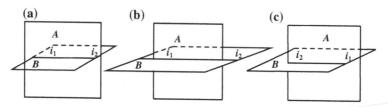

Fig. 7 a B passes through the plane segment A. **b** A passes through the plane segment B. **c** The planes partially cut each other

In Fig. 7a, *B* passes through A resulting in two intersection points within the boundary of *A*, whereas in Fig. 7b, *A* passes through *B* and results two intersection points within the boundary of *B*. In Fig. 7c, they cut partially; hence, one of the intersection points is within the boundary of *B* and the other one is within the boundary of *A*.

As a planar surface is always adjacent to either a spherical or a cylindrical surface, a partial cut refers to the fact that there may be intersection between the plane segments that partially cut and the adjacent surface of the other plane, i.e., one plane segment may cut the other in such a way that it cuts its neighboring cylindrical or spherical segment as well. In each of the cases, the planar–planar intersection is a line segment. But there is a difference in the above three cases; the first one has the intersection line segment starting from one point on the plane segment and ending at another point on the same plane segment having the intersection line segment fully on the plane segment, the second one has the intersection line segment beyond the plane segment, and in the third case the intersection line segment starts at a point on the plane segment while ending at a point beyond the plane segment. We discuss all the three cases below.

We define each plane segment by the equation of the plane and its boundary line segments, and therefore, the equations of lines. Now, we consider a plane and check for intersection between the line segments of the other planes. If we consider plane *A*, i.e., its equation and take the line segments of plane *B*, it results in i_1 and i_2. The points may either reside on or within the boundary of *A*. For each point, we traverse the boundary of *A* clockwise (or anti-clockwise) if the point is always on the right (or left) side, the point is within the boundary; otherwise, if the point satisfies one of the boundary line segments of *A*, the intersection point is on the boundary of the plane.

Planar—Spherical Intersection Whenever a plane cuts a spherical region once, the shape of the intersection is of a circular curve [10, 11]. Now, if the plane intersects a spherical region it results a circle, as the intersection curve has been depicted in Fig. 8a. As a spherical surface is adjacent to a set of cylindrical and planar surfaces, we may conclude that the planar surfaces that cut the spherical

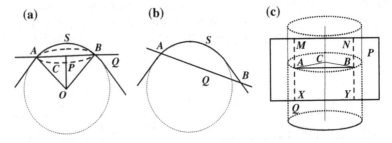

Fig. 8 The surface *Q* intersects the spherical surface *S*. **a** It results a circle *C* as the intersection *curve*. **b** When the surface (*Q*) partially intersects the spherical surface (*S*) that results a circular *curve* which is extended through the junction of the spherical and cylindrical or planar surfaces. **c** When a plane *P* intersects a cylinder *Q* resulting in two intersection line segments *MX* and *NY*

surface once, cut one or more adjacent surfaces as well, as shown in Fig. 8b, and has been discussed previously in case of planar–planar intersection.

Let S and Q be the spherical and planar surfaces. In Fig. 8a, a circle (C) is produced after the plane cuts the spherical surface. We draw a perpendicular OP on the plane. As we obtain the point P and the length of OP, we find the length of AP. Thus, the center (P) and the radius (AP) are known to us, and hence, we get the equation of the circle. Now, it may so happen that the plane does not cut the spherical surface at all. In that case, we check OP with the radius of the sphere. If OP is greater than the radius, there is no intersection.

Planar—Cylindrical Intersection When a plane segment cuts a cylinder, the intersection is a line segment satisfying both the equations of the cylinder and the plane [10, 11]. If the plane cuts the cylinder twice, we obtain two such line segments which are parallel to the axis of the cylinder. When the plane cuts the cylinder once, it means that there may be intersection with the adjacent surfaces of the cylindrical surface.

Let the plane P cut the cylinder Q twice, as shown in Fig. 8c. Then, we have to find two line segments MX and NY. We have the equation of cylinder as $(ny\text{-}mz)^2 + (lz\text{-}nx)^2 + (mx\text{-}ly)^2 = r^2(l^2 + m^2 + n^2)$ and its axis as $(x/l) = (y/m) = (z/n)$. Also the equation of the plane is $ax + by + cz = d$. We draw a perpendicular CO from C, a point on the axis, on the plane. As we know, the radius of the cylinder $(CA$ or $CB)$, AO (BO) can be directly found. Hence, we find the coordinate of point A. As the intersecting line segment passes through A and is parallel to the axis, we can derive its equation.

After knowing the equation, we check for intersection between the boundary line segments of planar surface, and obtain two intersection points on the boundary of the plane, here M and X. The line joining the two gives the line segment MX. Similarly, for point B, we get the line segment NY.

Spherical—Spherical Intersection If two spherical surfaces intersect, we obtain a circle as the intersecting curve and the equation of the circle satisfies equations of both the spheres [10, 11]. Let two spheres S and S' with their centers A and B, respectively, intersect with each other as depicted in Fig. 9a. If their equations are

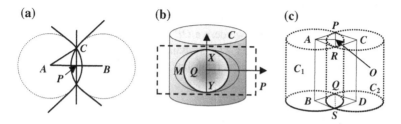

Fig. 9 a Two spheres with A and B as their centers intersect each other resulting in a circle as the intersection *curve* with P as the center and PC as the radius of the circle. **b** A sphere intersects the cylinder C resulting in M as the intersection curve in 3D, to find plane P that is drawn and the sphere cuts it along a circle Q. X and Y be the common points on M and Q. **c** Two cylinders C_1 and C_2 cut each other keeping their axes AB and CD parallel, results in two intersection line segments PQ and RS

$S \equiv x^2 + y^2 + z^2 + 2gx + 2fy + 2hz + c = 0$ and $S' \equiv x^2 + y^2 + z^2 + 2g'x + 2f'y + 2h'z + c' = 0$, the coordinate of the points of their intersection satisfying the equation $S–S' \equiv 2(g–g')x + 2(f–f')y + 2(h–h')z + (c–c') = 0$, which is the equation of the plane of intersection of the two spheres. This plane cuts either of the spheres in a circle. Our objective is to find out the circle thus obtained.

We draw a perpendicular from the center of one of the spheres on the plane. Here in the figure, the perpendicular AP is drawn on the plane. From the length of AP (as we can find the length of a perpendicular from a point outside a plane) and AC (the radius of the sphere), PC is calculated which is the radius of the intersecting circle. Again, as we know the coordinate of the center (P) and the radius (PC), the equation of the circle is immediately derived.

Spherical—Cylindrical Intersection When a spherical surface intersects a cylindrical surface, the intersecting curve does not lie in 2D plane [11]. It can be visualized by drawing a circle on a plane and then the plane is wrapped over the cylinder. The circle is not on the 2D plane now; rather, it is on the surface of the cylinder.

Here, in Fig. 9b, a sphere intersects a cylinder C resulting in the intersecting curve M. We have to derive the equation for M. At first, we draw a plane tangent to the cylinder, on which the line segment normal to the axis of the cylinder from the center of the sphere, is perpendicular. Now the sphere cuts the plane and results a circle Q as their intersecting curve. We derive the equation for this circle as we did in the planar-spherical intersection. Now, as we know the circle Q, we have its center and radius.

At this moment, to find M we need to know a point on it and then the locus of the point on the cylindrical surface. So, we draw a line passing through the center of the circle Q and parallel to the axis of the cylinder. This line cuts the circle at two points X and Y. The locus of either of the points satisfying the equation of the cylinder and maintaining the distance from the center of the circle as a constant provides the equation for M.

Cylindrical—Cylindrical Intersection If two cylinders intersect each other as in Fig. 9c, we obtain two intersecting line segments satisfying equations of both the cylinders. From the equations of the cylinders, we obtain the equation of the plane satisfying both the equations of the cylinders.

Let C_1 and C_2 be two cylinders. We draw AO, perpendicular on the intersecting plane from A. From AO and AP, we derive OP and hence obtain P. Also we find in the same way the point Q. Then PQ is attained as one of the intersecting line segments. Similarly, the other intersecting line segment RS is derived.

4 Computational Complexity

It is easy to observe that the guard zone of an n-vertex, ℓ-intersection line and p-plane convex solid object is a convex (3D) region with p planes, ℓ cylindrical arcs, and n spherical arcs only, when there is no intersection, and with intersection there might be p planes only, where $n = O(\ell)$ as well as $p = O(\ell)$. The planes of the guard zone are parallel to the planes of the solid object at a distance r apart, outside the solid object, and two adjacent planes of the guard zone are joined by a cylindrical arc of radius r centered at the associated intersection line of the solid object. Spherical arcs of radius r each are introduced as parts of the computed guard zone at the vertices of the solid object, where the associated planes of the guard zone are tangent to the cylindrical as well as spherical arcs and the cylindrical portions of the guard zone are also ending with spherical arcs of the guard zone near the vertices of the solid object. As a result, the time required for computing a 3D guard zone of a convex solid object is $O(n+\ell+p) = O(\ell)$.

Now, as per the next step, we need to draw the orthogonal projections of the individual guard zonal components, including planar, spherical, and cylindrical surfaces onto xy, yz, and zx planes. The projections can be drawn in linear time with respect to the number of guard zonal components, which is $O(\ell)$, where ℓ is the number of intersection lines present in the given 3D simple solid object. Therefore, for each guard zonal component, we obtain three different 2D objects at each of the xy, yz, and zx plane. Now, we need to merge the individual 2D components belonging to a specific 3D guard zonal component in such a way that the 2D components collectively form a 3D orthogonal box that encloses the corresponding 3D guard zonal component to its entirety. This step can be achieved in constant time and hence, the complexity of this step is $O(1)$.

Once the orthogonal boxes are ready, we feed these boxes into our customized space sweep algorithm for further processing. During the execution of the space sweep algorithm, the faces of each orthogonal boxes parallel to the xy, yz, and zx plane have been considered as event points. The event points are maintained in a dynamic list data structure, whereas the 3D boxes corresponding to the event points are maintained in a separate BST data structure for subsequent processing. The algorithm terminates when the event list becomes empty. Hence, the operations like creation, insertion, and deletion from the BST take $O(\ell \log \ell)$ time, whereas the similar operations for the dynamic list data structure consumes $O(\ell)$ time, for each of the axis planes. Therefore, the overall time complexity to perform this particular step requires $O(\ell \log \ell)$.

Now, let us assume that the execution of the previous step yields the following result, where the sweep plane xy has produced a set I containing all the overlapped box pairs while sweeping; afterward the yz plane operates on the pair of boxes contained in set I and produces a reduced set named J. Similarly, zx plane operates on the pair of boxes contained in set J and yields the final set K containing all the possible boxes that needs further investigation. Thus, xy $\rightarrow I$, yz $\rightarrow J$, and zx $\rightarrow K$.

All the operations explained above can be executed in $O(\ell \log \ell) + O(I \log I) + O(J \log J) \equiv O(\ell \log \ell + I \log I)$ time, since I dominates J.

As per the final step, we are left with the detection of intersection among the guard zonal components contained within the 3D boxes registered in set K. This operation can be executed in constant time for each pair of such guard zonal component, and hence, the time complexity for the intersection detection among the guard zonal components is $O(K)$.

The overall time complexity for the 3D guard zone computation for a given simple object is $O(\ell) + O(\ell \log \ell) + O(K) \equiv O(\ell \log \ell + I \log I)$.

5 Application

The guard zone computation problem occupies vast place of interest in the field of VLSI physical design automation and design of embedded systems, robotic motion control, geographic information system, etc. In VLSI physical design for optimized placement on the chip, we have to consider 3D subcircuits and their 3D guard zone for avoiding any parasitic effect. In this context, we deal each 3D subcircuit as a 3D simple solid object and accordingly we compute its guard zone. As cited for its 2D variation, we may achieve the goal in 3D counterpart.

In Robotics, it is important to have the motion planning feature built within the robot itself. This motion planning feature ensures that the robot does not collide with any obstacles while it is in motion unless it is programmed to do so. As we have noticed that Minkowski sum finds a tremendous application in motion planning of an object among obstacles [3], the guard zone computation can also be used to solve similar problems. Once a robot identifies the obstacles that it needed to bypass and transforms them into 2D simple polygons, then it can use the computed guard zone to avoid any possible collision. As the robot moves along its way, it keeps checking whether it encounters the already computed guard zone of any of the obstacles and if it finds one, then immediately it changes its direction unless it reaches to its destination. This process can further be enhanced by incorporating a learning mechanism within the robot where the robot records the objects found as an obstacle so far along with the computed guard zone and later when the robot encounters the similar objects, then it applies the already computed guard zone to avoid any possible collisions.

3D guard zone computation plays a significant role in robot motion planning, as the real life scenario conveys that the robots and obstacles faced are 3D in nature. Considering this fact it is obvious to take into account the 3D guard zone computing for efficient motion planning for 3D robots. Now, the problem of motion planning for robots can further be simplified if the robot somehow acquires the information of all the obstacles it is going to encounter from its start to final destination. Then the robot can compute its own simplified guard zone along with the guard zone for all the obstacles that it is going to encounter and then determines the path it is supposed to take to reach its destination. Another important application of the 3D guard zone computation problem is in the medical field. Consider the treatment of cancer cells in human body; now the biggest unsolved mystery till date is finding a targeted treatment for the cancer affected cells. There can be two different approaches of applying the guard zone computation in this regard: (1) A guard zone defining the growth of cancer affected cells in human body, definitely the medicine that should target these cells should have a faster healing power compared to the growth of the cancerous cells and also the medicines should have a predefined minimum guard zone which completely encompasses the affected cancerous cells. (2) A comparative study (with respect to the guard zonal effects) of various medicines that target the cancerous cells to determine which medicine needs to be applied on human body to reduce the effect of medicine on the healthy cells.

It also finds application in computing the buffer zone in geographical information systems [1], to name only a few.

6 Conclusion

As discussed earlier, resizing of electrical circuits is an important problem in VLSI layout design as well as in embedded system design, while accommodating the (groups of) circuit components on a chip floor. This problem motivates us to compute a guard zone of a simple polygon. In robot motion planning, geographic information system, embedded system 3D guard zone computation takes an important role. In this paper, we have considered the problem of computing a guard zone of a (3D) simple polygon, and developed a sequential algorithm for computing the same that uses the concepts of analytical and coordinate geometry to detect overlapped region(s) within the guard zone (if any) and accordingly exclude that region to report the resulting outer guard zone. Our algorithm can easily be modified to compute the regions of width r (as guard zonal distance) outside the polygon, and also inside the polygon (if necessary), which may find several applications in practice. This work can also be extended for computing a guard zone of a three-dimensional solid object that may not be a simple one, as a problem of probable future work.

References

1. Heywood, I., Cornelius, S., Carver, S.: An Introduction to Geographical Information Systems. Addison Wesley Longman, New York (1998)
2. Pottmann, H., Wallner, J.: Computational Line Geometry. Springer, Berlin (1997)
3. Lee, I.-K., Kimand, M.-S., Elber, G.: Polynomial/rational approximation of minkowski sum boundary curves (Article No.: IP970464). Graph. Models Image Process. **60**(2), 136–165 (1998)
4. Mehlhorn, K.: Data Structures and Algorithms 3: Multi-Dimensional Searching and Computational Geometry. Springer (1984); Bajaj, C., Kim, M.-S.: Generation of configuration space obstacles: the case of a moving algebraic curves. Algorithmica **4**(2), 157–172 (1989)
5. Chazelle, B.: Triangulating a simple polygon in linear time. Discrete Comput. Geom. **6**, 485–524 (1991)
6. Mehera, R., Chatterjee, S., Pal, R.K., A time-optimal algorithm for guard zone problem. In: Proceedings of 22nd IEEE Region 10 International Conference on Intelligent Information Communication Technologies for Better Human Life (IEEE TENCON 2007), CD: Session: ThCP-P.2 (Computing) (Four pages). Taipei (2007)

7. Mehera, R., Chatterjee, S., Pal, R.K.: Yet another linear time algorithm for guard zone problem. Icfai J. Comput. Sci. **II**(3), 14–23 (2008)
8. Mehera, R., Chakraborty, A., Datta, P., Pal, R.K.: An innovative approach towards detection and exclusion of overlapped regions in guard zone computation. In: Proceedings of 3rd International Conference on Computer, Communication, Control and Information Technology (C3IT 2015), pp. 1–6. Academy of Technology, West Bengal (2015)
9. Hwang, K., Briggs, F.A.: Computer Architecture and Parallel Processing. McGraw-Hill, New York (1984)
10. Grewal, B.S.: Higher Engineering Mathematics, 39th edn. Khanna Publishers, Delhi India (2005). ISBN 81-7409-195-5
11. Chakravorty, J.G., Ghosh, P.R.: Analytical Geometry and Vector Analysis. U.N. Dhur and Sons Private Ltd., Kolkata (2009)

Multistep Ahead Groundwater Level Time-Series Forecasting Using Gaussian Process Regression and ANFIS

N. Sujay Raghavendra and Paresh Chandra Deka

Abstract Groundwater level is regarded as an environmental indicator to quantify groundwater resources and their exploitation. In general, groundwater systems are characterized by complex and nonlinear features. Gaussian Process Regression (GPR) approach is employed in the present study to investigate its applicability in probabilistic forecasting of monthly groundwater level fluctuations at two shallow unconfined aquifers located in the Kumaradhara river basin near Sullia Taluk, India. A series of monthly groundwater level observations monitored during the period 2000–2013 is utilized for the simulation. Univariate time-series GPR and Adaptive Neuro Fuzzy Inference System (ANFIS) models are simulated and applied for multistep lead time forecasting of groundwater levels. Individual performance of the GPR and ANFIS models are comparatively evaluated using various statistical indices. In overall, simulation results reveal that GPR model provided reasonably accurate predictions than that of ANFIS during both training and testing phases. Thus, an effective GPR model is found to generate more precise probabilistic forecasts of groundwater levels.

Keywords ANFIS · Groundwater system · Gaussian process regression · Time-series forecasting

1 Introduction

Over the past decade, groundwater depletion is one of the major issues worldwide, which is posing direct or indirect impacts on human livelihoods, flora and fauna, natural habitat, and ecosystems. Depletion of groundwater storage, land subsidence,

N.S. Raghavendra (✉) · P.C. Deka
Department of Applied Mechanics and Hydraulics, National Institute
of Technology Karnataka, Surathkal, Mangalore 575025, India
e-mail: sujayraghavendran@ymail.com

P.C. Deka
e-mail: pareshdeka@yahoo.com

© Springer India 2016
R. Chaki et al. (eds.), *Advanced Computing and Systems for Security*,
Advances in Intelligent Systems and Computing 396,
DOI 10.1007/978-81-322-2653-6_19

reductions in stream flow and lake water levels, saltwater intrusion, loss of wetland and riparian ecosystems, and variations in groundwater quality are some of the vital factors influencing the sustainability of groundwater resources [1]. Sustainable groundwater resources development is the key issue to be addressed by policy makers or water managers by implementing various alternative management strategies. Groundwater restoration or recycle is not equally fast as that of surface water; it may take place after many years. Thus, constant monitoring of groundwater levels is extremely important for reliable assessment of temporal availability of groundwater at any required location [2]. The benefits of groundwater level forecasting include assessment of annual and long-term changes in groundwater storage, estimation of recharge rates, manage drinking water demand, and to ensure the sustainable use of groundwater resources [3].

Till date, several deterministic, stochastic and time-series based models have been developed for the forecasting of groundwater levels [4–7]. In the recent past, soft computing tools like Artificial Neural Network (ANN), ANFIS, Support Vector Regression (SVR), and so on have also been widely utilized for groundwater level prediction studies [8–14]. Quite a few hybrid artificial intelligence models developed by incorporating wavelet analysis efficiently forecast groundwater levels at different time scales [15–18]. Determining a model which is capable to efficiently capture the nonlinearities of the data without overfitting is the crucial job while modeling using time-series data. The ability to select the hyper parameters of the kernel automatically is one of the prominent benefits of Gaussian processes over conventional kernel interpretations of regression. The Bayesian learning algorithm-based Gaussian Process Regression is successfully applied for prediction of nonstationary time-series [19], monthly stream flow forecasting [20], and stream water temperature prediction [21], and so on. Compared to the conventional time-series forecasting methods, GPR model is said to possess strong nonlinear mapping ability, estimation of uncertainty, and is greatly fault-tolerant [22, 23]. Hence, in this paper, we demonstrate the state-of-the-art capability of Gaussian process regression for multistep lead time probabilistic forecast of groundwater level fluctuations. The ANFIS model is also employed for comparative study with GPR forecasts.

2 Study Area and Data Analysis

The study area (Fig. 1) is located near to southwest coast in the state of Karnataka, India. The observation wells selected for the current study are located inside the Kumaradhara river basin which covers a geographical area of 1776 sq km and is located in between 12° 29′ 04″ and 12° 58′ 33″ north latitude and from 75° 09′ 58″ to 75° 47′ 48″ east longitudes. The observation well located at Bellare lies at 12° 39′ 53″ north latitude and 75° 17′ 18″ east longitude, while the other well at Guttigaru lies at 12° 37′ 53″ north latitude and 75° 31′ 44″ east longitude as shown in Fig. 1.

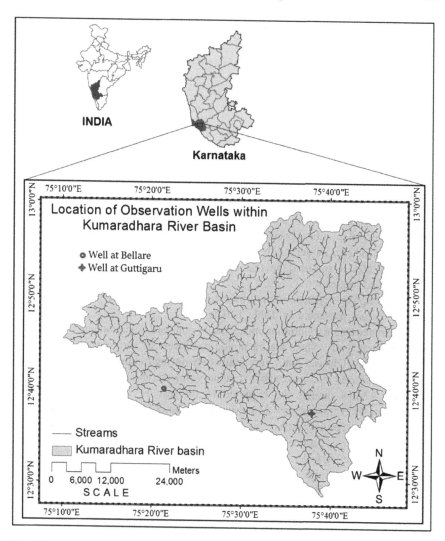

Fig. 1 Study area (location of observation wells)

The study area has a tropical monsoon climate dominated by the southwest monsoon (June–October). The mean annual rainfall over the basin is around 3,500 mm. The geology of the area is predominantly characterized by Lateritic soil with highly porous and permeable nature. Due to this lateritic soil property, shallow groundwater levels in the selected unconfined aquifers follow a regular cyclic pattern of seasonal fluctuation, typically rising during the monsoon due to greater precipitation and recharge, then declining during the summer.

The groundwater level data of the observation wells located at Bellare and Guttigaru for the years 2000–2013 were retrieved from Department of Mines

Table 1 Statistical parameters of groundwater level dataset

Groundwater table below ground level (m)				
Well location	Bellare		Guttigaru	
Dataset	Train	Test	Train	Test
X_{max}	13.05	12.13	11.24	12.38
X_{min}	4.28	5.21	1.33	3.2
X_{mean}	9.25	8.53	7.11	7.37
X_{mode}	10.18	5.58	4.27	3.29
S_d	2.25	2.26	2.98	3.15
C_v	0.24	0.27	0.42	0.43

and Geology, Dakshina Kannada Dist., Govt. of Karnataka, India. The topographic elevation of these wells is about 100–130 m above mean sea level. This data comprises of 166 monthly observations, in which 10 years of data during Jan 2000–Dec 2009 is used for model training and remaining 4 years of data during Jan 2010–Oct 2013 is used as out-of-sample set or testing set to measure the predictability of the developed models.

The descriptive statistics of the observed groundwater levels in the two observation wells are presented in Table 1. The X_{max}, X_{min}, X_{mean}, X_{mode}, S_d, and C_v denotes the maximum, minimum, mean, mode, standard deviation and coefficient of variation respectively. Since the mean and coefficient of variation of the observed groundwater level dataset don't vary ominously during training and testing periods, it could be inferred as a reasonable stationary time-series. In the present scenario, GPR and ANFIS models are explored to forecast 1, 3 and 6 months ahead groundwater level fluctuations. Monthly groundwater level time-series up to previous four time steps are taken as input variables. In order to test the hypothesis that $GWL_{(t-2)},...GWL_{(t-p)}$ further help in forecasting $GWL_{(t)}$, beyond $GWL_{(t-1)}$, one can use an F-test. The lag order $p = 4$ was determined from the F-test statistic. F-test is the test statistic to examine the significance of the components in the model [24]. The expected output from the developed models is the groundwater level at time step t, t + 3, and t + 6. The input-output combinations are as presented below.

I. $GWL_{(t-4)} + GWL_{(t-3)} + GWL_{(t-2)} + GWL_{(t-1)} = GWL_{(t)}$

II. $GWL_{(t-4)} + GWL_{(t-3)} + GWL_{(t-2)} + GWL_{(t-1)} = GWL_{(t+3)}$

III. $GWL_{(t-4)} + GWL_{(t-3)} + GWL_{(t-2)} + GWL_{(t-1)} = GWL_{(t+6)}$

3 Methodology

In the present study, Gaussian Process Regression (GPR) and Adaptive Neuro Fuzzy Inference System (ANFIS) approaches are proposed for model development of groundwater level time-series forecasting. GPR and ANFIS is used for 1, 3, and 6 month lead groundwater level time-series forecasting using lagged input data up to 4 months in the past.

3.1 Gaussian Process Regression

Gaussian process regression is a standard method in probability theory wherein the interpolated values are modeled by a Gaussian process governed by prior covariance. Incorporating appropriate assumptions on the priors, GPR renders the best linear unbiased prediction of the values [25]. GPs constitute one of the most important Bayesian discriminative kernel learning approach due to its practical and theoretical simplicity and outstanding generalization ability. A sequence of random variables $\{X_n\}$ defining a stationary process can have any probability distribution. A stationary process $\{X_n\}$ is called a Gaussian process, if the joint distribution of $(X_{n+1}, X_{n+2},..., X_{n+k})$ is a k-variate normal for every positive integer k.

Consider an observation space χ. A GP $f(x)$, where $x \in \chi$, is defined by a set of random variables, any finite number of which possess a joint Gaussian distribution function which is fully specified by its mean function $m(x)$ and covariance $k(x, x')$ [26].

So let,

$$m(x) = E[f(x)]$$
$$k(x, x') = E[(f(x) - m(x)) \cdot (f(x') - m(x))] \tag{1}$$

Now we can write GP as

$$f(x) \sim \mathbb{N}(m(x), k(x, x')) \tag{2}$$

Consider a training set $D = \{(x_i, y_i) | i = 1, 2, ..., N\}$, with m-dimensional input variables, x_i being the observed data related to the phenomenon that is to be modeled and scalars y_i being the associated target values given by $y_i = f(x_i) + \epsilon_i$, where ϵ_i is Gaussian noise with zero mean and variance σ_n^2.

The joint normality of the training target values $y = [y_i]_{i=1}^N$ and some unknown target value y_*, are estimated by the value f_* of the hypothesized GP assessed at the observation point x_*, yields

$$\begin{bmatrix} y \\ f_* \end{bmatrix} \sim \mathbb{N}\left(0, \begin{bmatrix} K(X, X) + \sigma_N^2 I_N & k(x_*) \\ k(x_*)^\mathrm{T} & k(x_*, x_*) \end{bmatrix}\right) \tag{3}$$

where,

$$k(x_*) \triangleq [k(x_1, x_*), ..., k(x_N, x_*)]^T \tag{4}$$

$X = [x_i]_{i=1}^N$, I_N, is the $N \times N$ identity matrix, $k(x_*)$ is the vector of covariance between f_* and the training latent function values, and K is the matrix of the covariance between the N training data points (design matrix)

$$K[X,X] \triangleq \begin{bmatrix} k(x_1,x_1) & k(x_1,x_2) & \ldots\ldots & k(x_1,x_N) \\ k(x_2,x_1) & k(x_2,x_2) & \ldots\ldots & k(x_2,x_N) \\ & & & \\ \cdot & \cdot & & \cdot \\ \cdot & \cdot & & \cdot \\ k(x_N,x_1) & k(x_N,x_2) & \ldots\ldots & k(x_N,x_N) \end{bmatrix} \tag{5}$$

Then, from (Eq. 3) and conditioning on the available training samples, we can derive the expression for the model predictive distribution, yielding

$$p(f_*|x_*, D) = \mathbb{N}(f_*|\mu_*, \sigma_*^2), \tag{6}$$

where

$$\begin{cases} \mu_* = k(x_*)^{\mathrm{T}} \left(K(X,X) + \sigma_N^2 I_N \right)^{-1} \times y \\ \sigma_*^2 = \sigma_N^2 - k(x_*)^{\mathrm{T}} \left(K(X,X) + \sigma_N^2 I_N \right)^{-1} k(x_*) + k(x_*, x_*) \end{cases} \tag{7}$$

The covariance function is parameterized by optimal value of hyper parameters. The predictive variance of the GP model is as given in Eq. (7), and it does not depend on the training target values, but depends only on the training input values [27]. The optimal value of hyper parameters of a Gaussian process with any kernel θ, for any distinct data set can be derived by maximizing the log marginal likelihood by means of general optimization procedures. The log marginal likelihood function under the GPR model is presented in Eq. 8 given below.

$$\begin{cases} \log p(y|X; \theta, \sigma^2) = -\dfrac{N}{2} \log 2\pi - \dfrac{1}{2} \log \left| K(X,X) + \sigma_N^2 I_N \right| \\ \qquad\qquad\qquad - \dfrac{1}{2} y^{\mathrm{T}} \left(K(X,X) + \sigma_N^2 I_N \right)^{-1} y \end{cases} \tag{8}$$

3.2 Adaptive Neuro Fuzzy Inference System (ANFIS)

ANFIS is the fuzzy-logic based paradigm integrated with the learning power of Artificial Neural Network (ANN) to improve the intelligent system's performance utilizing knowledge acquired after learning. For a given input--output data set, ANFIS constructs a hybrid learning algorithm that associates the backpropagation gradient descent and least squares methods to frame a fuzzy inference system whose membership function (MF) parameters are iteratively tuned or adjusted. Adaptive Neuro Fuzzy inference systems comprise of mainly five layers--rule base, database, fuzzification interface, defuzzification interface and decision-making unit. The generalized ANFIS architecture proposed by Jang [28] is summarized below (Fig. 2).

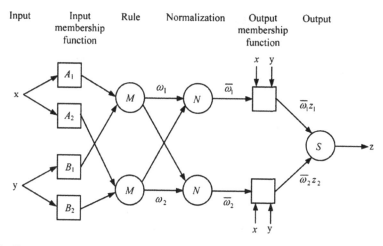

Fig. 2 General ANFIS architecture with two membership functions on each of the two inputs

ANFIS architecture comprises of five layers. Every single node in layer 1 is an adaptive node with a node function which may be anyone among the membership functions. Every node of layer 2 is a fixed node labeled 'M' which signposts the firing strength of each rule. All nodes of layer 3 are fixed nodes labeled as 'N' which demonstrates the normalized firing strength of each rule. The Layer 4 is as similar to layer 1 wherein every node is an adaptive node governed by a node function. The layer 5 being a single fixed node labeled 'S', representing the overall output (z), defined as the summation of all incoming signals [29].

In the present study, we examine three types of membership functions (MFs) namely trapezoidal, gaussian, and generalized bell. Among all the three types of the MFs, we impart two MFs on each of our four inputs, in which eight altogether. With this, the FIS structure consists of 16 fuzzy rules with 104 parameters. A hybrid algorithm integrating the least squares method and the backpropagation gradient descent method is applied to optimize and adjust the generalized bell membership function parameters and coefficients of the output linear equations. The number of epochs and error tolerance is set to 1000 and 0, respectively. From the result as presented in Table 3, it is determined that the ANFIS structure with Generalized bell MF to be better performing than Trapezoidal and Gaussian shaped MFs based on the performance evaluation using correlation coefficient statistic as mentioned below in Sect. 4. Hence, generalized bell MF-based ANFIS models are developed for all the 1, 3 and 6 month lead time forecasting scenarios.

4   Performance Evaluation

The following statistical indices are used to evaluate the performance of both the GPR and ANFIS models in forecasting groundwater level time-series.

$$CC = \frac{\sum\limits_{i=1}^{N}\left\{(X_i - \overline{X}) \cdot (Y_i - \overline{Y})\right\}}{\sqrt{\sum\limits_{i-1}^{N}\left\{(X_i - \overline{X})^2 \cdot (Y_i - \overline{Y})^2\right\}}} \tag{9}$$

$$RMSE = \sqrt{\frac{\sum\limits_{i=1}^{N}(X_i - Y_i)}{N}} \tag{10}$$

$$NSE = 1 - \frac{\sum\limits_{i=1}^{N}(X_i - Y_i)^2}{\sum\limits_{i=1}^{N}(X_i - \overline{X})^2} \tag{11}$$

where,
CC   Correlation Coefficient;
RMSE   Root Mean Squared Error;
NSE   Nash-Sutcliffe Efficiency;
X   Observed/Actual values;
Y   Modeled/Computed values;
$\overline{X}$ Mean of Actual data.

5   Results and Discussion

An appealing characteristic of time-series modeling is that it is based on relatively few assumptions which usually lead to yield good fits. The GPR package in the WEKA 3.6 software [30] is employed to develop the GPR models. The GPR employing Pearson VII function-based universal kernel (PuK) is used for model development. The GPR model developed in the present study is propelled to provide better groundwater level forecasting results. Table 2 presents the developed GP regression model equations. The statistical adequacies of the GPR and ANFIS models for 1, 3 and 6 month ahead forecasts are summarized in Tables 4, 5, and 6, respectively. For both study sites (Bellare and Guttigaru), the GPR models are found to provide more accurate groundwater level forecasts than that of ANFIS

Table 2 Values of Gaussian process regression equations

GPR forecast	Average target value	Inverted covariance matrix		Inverted covariance matrix × target-value vector	
		Lowest value	Highest value	Lowest value	Highest value
Groundwater monitoring well near Bellare					
1 Month lead	9.1686	−0.2107	0.9420	−4.2888	2.1221
3 Month lead	9.1978	−0.2107	0.9420	−4.6319	3.5733
6 Month lead	9.2933	−0.2107	0.9420	−5.0088	3.6627
Groundwater monitoring well near Guttigaru					
1 Month lead	7.0066	−0.1949	0.9179	−4.4780	3.4013
3 Month lead	7.0806	−0.1949	0.9179	−3.6057	5.1893
6 Month lead	7.2328	−0.1949	0.9179	−4.3094	4.5839

Table 3 Performance of ANFIS models at 1 month lead time forecasting

Correlation coefficient (CC)				
ANFIS model with	Bellare		Guttigaru	
	Train	Test	Train	Test
Trapezoidal MF	0.6	0.54	0.58	0.55
Gaussian MF	0.81	0.74	0.79	0.71
Generalized Bell MF	0.95	0.85	0.89	0.83

models for 1, 3 and 6 month lead time forecasting. The GPR models for the Bellare and Guttigaru well sites have a testing RMSE of 0.632 and 1.05 m, respectively (Table 4), and are superior to the ANFIS model forecast, which has a testing RMSE of 0.742 m for the Bellare well site and 1.39 m for the Guttigaru well site during 1 month lead time forecasting (Table 3).

It can be observed (from Tables 4, 5 and 6) that the correlation coefficients of both the GPR and ANFIS models are high during training (calibration). However, during the testing phase, the GPR model is better when compared to ANFIS model. It is noteworthy that the GPR model shows enhanced performance in contrast to ANFIS model, in case of both the wells. The RMSE statistic of multistep lead time forecasting is presented in Fig. 3 wherein it can be inferred that the GPR and ANFIS models are more capable in the shorter lead time forecast. It can be seen that the forecasting efficiency declines during longer lead time forecast. The ANFIS model performs marginally similar to GPR model for 1 month ahead groundwater level forecasting, but for the higher lead times, such as 3 and 6 month lead time, GPR performance is observed better than ANFIS model results as presented in Tables 4, 5 and 6.

Table 4 Performance of GPR and ANFIS models during 1 month lead time forecasting

Well location		Bellare			Guttigaru		
Statistical indices		RMSE (m)	NSE	CC	RMSE (m)	NSE	CC
GPR	TRAIN	0.577	0.93	0.97	0.97	0.89	0.94
	TEST	0.632	0.92	0.94	1.05	0.87	0.92
ANFIS	TRAIN	0.66	0.91	0.95	1.32	0.84	0.89
	TEST	0.742	0.82	0.85	1.39	0.8	0.83

Table 5 Performance of GPR and ANFIS models during 3 month lead time forecasting

Well location		Bellare			Guttigaru		
Statistical indices		RMSE (m)	NSE	CC	RMSE (m)	NSE	CC
GPR	TRAIN	0.74	0.89	0.91	1.105	0.86	0.91
	TEST	0.82	0.86	0.89	1.211	0.83	0.88
ANFIS	TRAIN	0.89	0.84	0.88	1.69	0.81	0.86
	TEST	1.09	0.79	0.84	1.847	0.76	0.82

Table 6 Performance of GPR and ANFIS models during 6 month lead time forecasting

Well location		Bellare			Guttigaru		
Statistical indices		RMSE (m)	NSE	CC	RMSE (m)	NSE	CC
GPR	TRAIN	0.96	0.83	0.87	1.225	0.83	0.85
	TEST	1.167	0.78	0.81	1.37	0.80	0.82
ANFIS	TRAIN	1.281	0.79	0.82	1.82	0.76	0.8
	TEST	1.41	0.74	0.78	1.97	0.71	0.75

Fig. 3 RMSE of GPR and ANFIS models at multistep lead time forecasting

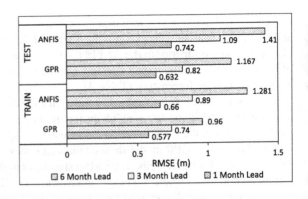

Figures 4 and 5 illustrate observed versus forecasted groundwater level time-series using GPR and ANFIS models. It can be seen from Figs. 4 and 5 that the GPR model can efficiently mimic observed groundwater level time-series better than ANFIS model during 1 month lead forecasting. Figures 6 and 7 are scatter

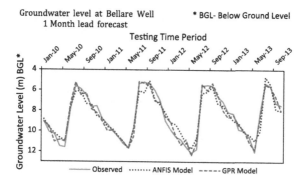

Fig. 4 Plot of observed versus forecasted groundwater level time-series with respect to the well location at Bellare of 1 month lead time forecasting models

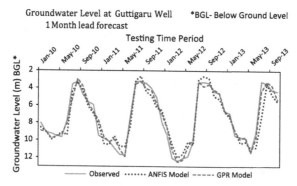

Fig. 5 Plot of observed versus forecasted groundwater level time-series with respect to well location at Guttigaru of 1 month lead time forecasting models

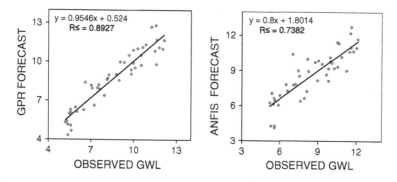

Fig. 6 Scatter plot of observed versus forecasted groundwater level with respect to well the location at Bellare of 1 month lead time forecasting models during test phase

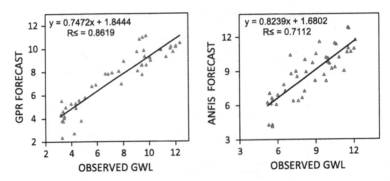

Fig. 7 Scatter plot of observed versus forecasted groundwater level with respect to well location at Guttigaru of 1 month lead time forecasting models during test phase

plots comparing the observed and forecasted groundwater levels using the GPR and ANFIS models for 1 month lead time forecasting during the testing period at the Bellare and Guttigaru sites. It can be observed that the band of scatter plot is very narrow and close to the line of perfect fit in case of GPR forecast, On the other hand ANFIS shows marginally lesser performance as compared to the GPR model in test phase. On a whole, it can be concluded that the GPR model provided more accurate forecasting results at both the study sites than the best ANFIS model at all the 1, 3 and 6 month lead times considered.

6 Conclusions

The application of the Gaussian Process Regression to forecast monthly ground-water level fluctuations at multistep lead times is investigated in the present study. ANFIS modeling is also adopted for comparative performance evaluation of the developed models. It is observed that the performance of the GPR is quite satis-factory providing relatively close agreement predictions when compared to that of ANFIS model in terms of the performance measures utilized in this study. It is envisaged that GPR model could serve as a better alternate for forecasting groundwater level fluctuation at multistep lead time. The GPR model has advan-tages over other models in terms of model accuracy, feature scaling, and proba-bilistic variance. In future one can test the applicability of GPR model with multivariate input data to forecast groundwater levels by including rainfall, tem-perature, and evaporation data.

Acknowledgements The authors would like to thank the Department of Mines and Geology, Government of Karnataka for providing the necessary data required for research and the Department of Applied Mechanics and Hydraulics, National Institute of Technology Karnataka for the necessary infrastructural support. The authors would like to thank four anonymous reviewers for their valuable suggestions and comments.

References

1. Alley, W.M., Reilly, T.E., Franke, O.L.: Sustainability of Ground-Water Resources, p. 79. U. S. Geological Survey Circular 1186, Denver (1999)
2. Raghavendra, N.S., Deka, P.C.: Sustainable development and management of groundwater resources in mining affected areas. Procedia Earth Planet. Sci. **11**, 598–604 (2015). doi:10. 1016/j.proeps.2015.06.061
3. Taylor, C.J., Alley, W.M.: Ground-Water-Level Monitoring and the Importance of Long-Term Water-Level Data, p. 67. U.S. Geological Survey Circular 1217, Denver (2001)
4. Gupta, A.D., Onta, P.R.: Sustainable groundwater resources development. Hydrol. Sci. J. **42**, 565–582 (1997)
5. Adamowski, K., Hamory, T.: A stochastic systems model of groundwater level fluctuations. J. Hydrol. **62**, 129–141 (1983)
6. Ahn, H.: Modeling of groundwater heads based on second-order difference time series models. J. Hydrol. **234**, 82–94 (2000)
7. Bidwell, V.J.: Realistic forecasting of groundwater level, based on the eigenstructure of aquifer dynamics. Math. Comput. Simul. 12–20 (2005)
8. Sudheer, Ch., Mathur, S.: Groundwater level forecasting using SVM-PSO. Int. J. Hydrol. Sci. Technol. **2**, 202 (2012)
9. Daliakopoulos, I.N., Coulibaly, P., Tsanis, I.K.: Ground water level forecasting using artificial neural networks. J. Hydrol. **309**, 229–240 (2005)
10. Shirmohammadi, B., Vafakhah, M., Moosavi, V., Moghaddamnia, A.: Application of several data-driven techniques for predicting groundwater level. Water Resour. Manag. **27**, 419–432 (2013)
11. Nourani, V., Ejlali, R.G., Alami, M.T.: Spatiotemporal Groundwater Level Forecasting in Coastal Aquifers by Hybrid Artificial Neural Network-Geostatistics Model: A Case Study (2011)
12. Raghavendra, N.S., Deka, P.C.: Forecasting monthly groundwater table fluctuations in coastal aquifers using Support vector regression. In: Anadinni, S. (ed.) International Multi Conference on Innovations in Engineering and Technology (IMCIET-2014), pp. 61–69. Elsevier Science and Technology, Bangalore (2014)
13. Shiri, J., Kisi, O., Yoon, H., Lee, K.-K., Nazemi, A.H.: Predicting groundwater level fluctuations with meteorological effect implications-{A} comparative study among soft computing techniques. Comput. Geosci. **56**, 32–44 (2013)
14. Yoon, H., Jun, S.-C., Hyun, Y., Bae, G.-O., Lee, K.-K.: A comparative study of artificial neural networks and support vector machines for predicting groundwater levels in a coastal aquifer. J. Hydrol. **396**, 128–138 (2011)
15. Suryanarayana, C., Sudheer, C., Mahammood, V., Panigrahi, B.K.: An integrated wavelet-support vector machine for groundwater level prediction in Visakhapatnam, India. Neurocomputing **145**, 324–335 (2014)
16. Adamowski, J., Chan, H.F.: A wavelet neural network conjunction model for groundwater level forecasting. J. Hydrol. **407**, 28–40 (2011)
17. Maheswaran, R., Khosa, R.: Long term forecasting of groundwater levels with evidence of non-stationary and nonlinear characteristics. Comput. Geosci. **52**, 422–436 (2013)
18. Raghavendra, N.S., Deka, P.C.: Forecasting monthly groundwater level fluctuations in coastal aquifers using hybrid Wavelet packet—Support vector regression. Cogent Eng. **2**, 999414 (2015)
19. Brahim-Belhouari, S., Bermak, A.: Gaussian process for nonstationary time series prediction (2004)
20. Sun, A.Y., Wang, D., Xu, X.: Monthly streamflow forecasting using Gaussian process regression. J. Hydrol. **511**, 72–81 (2014)
21. Grbić, R., Kurtagić, D., Slišković, D.: Stream water temperature prediction based on Gaussian process regression. Expert Syst. Appl. **40**, 7407–7414 (2013)

22. Roberts, S., Osborne, M., Ebden, M., Reece, S., Gibson, N., Aigrain, S.: Gaussian processes for time-series modelling. Philos. Trans. A. Math. Phys. Eng. Sci. **371**, 20110550 (2013)
23. Yan, W., Qiu, H., Xue, Y.: Gaussian process for long-term time-series forecasting. In: Proceedings of the International Joint Conference on Neural Networks, pp. 3420–3427 (2009)
24. Box, G.E.P.: Non-Normality and tests on variances. Biometrika **40**, 318–335 (1953)
25. Rasmussen, C.E.: Gaussian processes in machine learning. In: Advanced Lectures on Machine Learning. Lecture Notes in Computer Science: Lecture Notes in Artificial Intelligence, pp. 63–71. Springer, Germany (2004)
26. Mackay, D.J.C.: Introduction to Gaussian processes. Neural Netw. Mach. Learn. **168**, 133–165 (1998)
27. Rasmussen, C.E., Williams, C.: Gaussian processes for machine learning. Adaptive Computation and Machine Learning, p. 272. The MIT Press, Cambridge (2006)
28. Jang, J.S.R.: ANFIS: adaptive-network-based fuzzy inference system. IEEE Trans. Syst. Man Cybern. **23**, 665–685 (1993)
29. Keskin, M.E., Taylan, D., Terzi, Ö.: Adaptive neural-based fuzzy inference system (ANFIS) approach for modelling hydrological time series. Hydrol. Sci. J. **51**, 588–598 (2006)
30. Hall, M., National, H., Frank, E., Holmes, G., Pfahringer, B., Reutemann, P., Witten, I.H.: The WEKA data mining software: an update. SIGKDD Explor. **11**, 10–18 (2009)

Anomaly Detection and Three Anomalous Coins Problem

Arpan Chakraborty, Joydeb Ghosh, Piyali Datta, Ankita Nandy
and Rajat Kumar Pal

Abstract Counterfeit coin problem has been considered for a very long time and is a topic of great significance in Mathematics as well as in Computer Science. In this problem, out of n given coins, one or more false coins (the coins are classified as false because of their different weight from a standard coin) are present which have the same appearance as the other coins. The word *counterfeit* or *anomalous* means something deviated from the standard one. In this respect, finding out these *anomalous* objects from a given set of data items is of utmost importance in data learning problem. Thus, representing coins as any data items, we have introduced an algorithm to determine three false coins out of n given coins. In addition, our objective is to solve the problem in minimum number of comparisons with the help of an equal arm balance.

Keywords Counterfeit coin · Anomaly detection · Weighing · Arm balance · Decision tree · Algorithm · Complexity

A. Chakraborty (✉) · P. Datta · A. Nandy · R.K. Pal
Department of Computer Science and Engineering, University of Calcutta,
Acharya Prafulla Chandra Roy Siksha Prangan, JD – 2, Sector – III, Saltlake City,
Kolkata 700098, West Bengal, India
e-mail: arpanc250506@gmail.com

P. Datta
e-mail: piyalidatta150888@gmail.com

R.K. Pal
e-mail: pal.rajatk@gmail.com

J. Ghosh
Department of Mathematics, Surendra Institute of Engineering and Management,
New Chamta, Siliguri, Darjeeling 734009, West Bengal, India
e-mail: joydeb009@gmail.com

© Springer India 2016
R. Chaki et al. (eds.), *Advanced Computing and Systems for Security*,
Advances in Intelligent Systems and Computing 396,
DOI 10.1007/978-81-322-2653-6_20

303

1 Introduction

Counterfeit or anomaly detection problem is a very well-known domain that holds immense importance in the field of Mathematics, Computer Science as well as in security issue. Most frequently, we have to deal with a huge collection of data that are defined by some parameters. The parameters hold specific values depending on the problem instance and there may exist some data that deviate from these specific values; accordingly, we consider these data items to be anomalous. Now, the problem is to detect the counterfeited data by means of some testing mechanism of the items where we even do not know the actual value of the parameter of the standard item and the anomalous items. As this problem considers a number of times the testing method is used as its required cost, our objective is to minimize that number in order to minimize the overall cost.

Here n coins problem plays an important role. By representing the coins as the data item, its weight as the defining parameter and weighing of the coins using a single arm balance essentially formulate the anomaly detection problem as a whole. In this respect, our objective is to minimize the number of weighings for which it is sufficient to determine the defective coin(s) in a set of n coins using only an equal arm balance, when the number of odd coins is precisely known and they are identical in appearance but different in weight (either heavier or lighter) than a true coin. Beyond the theoretical realm, computing a solution of the counterfeit coin(s) problem has huge significance in commercial sphere as well as to prevent forgery in different fields.

Finding one counterfeit coin out of n coins is a complex problem. However, the complexity of the problem increases manifold when more than one counterfeit coin is present in a set of n coins. When only one counterfeit coin is there in a set of n coins and if it is given whether it is heavier or lighter than the true coins, using an equal arm balance our problem becomes a binary search problem. If the nature of the false coin is not mentioned, the false coin can either be heavier or lighter than the standard coins, and hence, it is necessary to proceed accordingly with the subset of coins to find out the false coin.

Unlike in the one counterfeit coin problem, the two counterfeit coins problem comprises several variations with respect to the relationship between the two false coins. The two false coins can be equally heavier (or equally lighter), unequally heavier (or unequally lighter) or the two false coins may be present in a combination of heavier and lighter (than the standard coin) coins. Hence, for the two counterfeit coins problem, a large sample space has to be taken care of. Here optimizing the number of comparisons to find out the false coin is the challenge as the complexity of the algorithm depends on the number of comparisons required. Now we take our attention to another level where we find three counterfeit coins in a set of n number of coins. As is evident, this problem becomes more complex than the two coins problem. However, we can use the results obtained from the two counterfeit coins problem while we construct algorithms for a variety of versions dealing with the presence of three false coins in a set of coins. As we will shortly see, there are a number of adaptations of the three coins problem.

2 Literature Survey

In articles [1, 2], one solution to the single counterfeit coin problem has been developed in the form of a decision tree that represents a set of all feasible decisions by which we can attain the desired solution(s) of the problem. In this solution, each internal vertex (that is not a leaf vertex) symbolizes an assessment between a pair of equal sets of coins using an equal arm balance. Here the problem under concern is more generalized; the forged coin can either be heavier or lighter than a true coin. So, for n given coins, there are $2n$ leaf vertices in the tree as plausible solutions.

In paper [3], the problem of finding the minimum number of weighings that suffices to determine the counterfeit (heavier) coins in a set of n coins of identical appearance, given a balance scale and the information that there are precisely two heavier coins at hand, has been considered. Both the heavier coins are of the same weight and they are not heavier than 1.5 times than a true coin. If P is the maximum number of comparisons mandatory to find out two false coins (equally heavier), the paper devises an algorithm that has the lower bound $\lceil \log 3(nC2) \rceil$. In this paper, a set of n, has been considered, for which this lower bound is achieved, and the upper bound that is only one unit more than the lower bound.

In paper [4], the problem has been introduced as a relevance of dynamic programming and the allied analysis has been performed through optimal and suboptimal testing policy. Here, the algorithm is developed considering the single false coin problem. This procedure takes an assumption that, $k < n$ coins are there in each pan for each weighing, where the value of k effectively depends on the value of n. If the two groups weigh equal, the defective coin must be in the left over $n - 2k$ coins; otherwise, the false coin is in one of the k groups. Whenever each weighing is over, the number of coins to be examined diminishes; however, the problem remains the same. This allows the authors to pertain dynamic programming to this problem.

In paper [5], the problem has been introduced in two ways. In the first case, it is not convinced whether there is a counterfeited coin in the given set (or not). The algorithm checks it first, and if any, then recognizes the fake coin by means of a minimum number of weighings. In the second case, it is known that there is a counterfeit coin and the intention is to find the coin through a minimum number of weighings. If required, a standard coin may also be provided. In the first case, if a lighter coin is there in the given set S of coins, then it is proved that the least number of weighings to find out the fake coin satisfies $3^{n-1} < |S| \leq 3^n$ for some distinctive value of n, where $|S|$ is the cardinality of set S. In the second case, we are given a set S of coins plus a standard coin, where only one coin in S is of dissimilar weight. Then it is proved that $(3^{n-1}-1)/2 < |S| \leq (3^n-1)/2$.

Let S be a set of more than two coins, out of which only one is a counterfeit coin. In this case, the least number of weighings to find out the forged coin satisfies $(3^{n-1}-3)/2 < |S| \leq (3^n-3)/2$, for some sole value of n. The paper asserts that a lighter coin can be found in $\lceil \log_3 k \rceil$ weighings, if there are k coins in the set. According to the paper, $\lceil \log_3(2k+1) \rceil$ is the complexity of the algorithm for solving the problem, if an additional standard coin is given; otherwise, the complexity is $\lceil \log_3(2k+3) \rceil$.

In articles [6–9], there are four algorithms to solve two counterfeit coins problem. For two coins problem, difficulty of the problem increases than that of single coin problem as all the false coin combinations have to be considered. Again, in paper [6], the authors have considered the case of finding two equally heavier (lighter) coins. Here decision tree has been used as the data structure and the main issue of the algorithm is to subdivide the coins into three nearly equal sets and through some comparisons of the specified sets the algorithm ultimately finds the false coins pair. In this paper, the main objective is to reduce the sample space by a factor of three after each comparison; hence, the time complexity of the algorithm becomes O(log n). The leaves of the decision tree show all possible false coin combinations.

3 Formulation of the Problem

We imagine that in the search space, there are n coins all of which are identical in appearance. By a standard or true coin we mean that its weight is specified to a value, say x unit, and a fake (or false or counterfeit) coin is a specified coin which only differs from a standard one with respect to weight. If the weight of a fake coin is y unit, then following situations arise: x > y; x < y, i.e., the fake coin is either lighter or heavier than a true coin. Now, it is mentioned that we do not have the weights of the coins. We are only provided with a single arm balance. By using the arm balance, i.e., by comparing the weights of the coins among themselves, we have to find out the fake coin(s) from the search space. Now depending on the number of fake coins in the search space, the problem becomes complex. Imagine a coin is false among 10 coins. We put 5 coins in the left pan and 5 coins in the right pan, and say, the left pan goes upside and right pan downside. From this weighing, we may assume two things: either the fake coin is heavier residing on the right pan and the coins on the left are all true, or the fake coin is lighter residing on the left pan and all the coins on the right are true. Thus, from this single weighing we cannot conclude anything else.

Let T, H, and L denote a true coin, a heavier false coin, and a lighter false coin, respectively. Their weights are denoted as w(T), w(H), and w(L), respectively. For one counterfeit coin problem as only single coin is fake, there are two versions:

$w(T) > w(L)$ and $w(T) < w (H)$. Now, if two coins are false, then there will be six possibilities:

1. $w(H_1) = w(H_2)$, i.e., both the false coins are equally heavier, where H_1 and H_2 are two heavier false coins.
2. $w(L_1) = w(L_2)$, i.e., both the false coins are equally lighter, where L_1 and L_2 are two lighter false coins.
3. Both the counterfeit coins are unequally heavier, i.e., $w(H_1) <> w(H_2)$.
4. $w(L_1) <> w(L_2)$, i.e., both the counterfeit coins are unequally lighter.
5. One of the false coins is heavier whereas another is lighter than a true coin. In this case, one of the variations takes place when the false coins are equally heavier and lighter, i.e., the difference in weight of the heavier coin and a standard coin is equal to the difference in weight of a standard coin and the lighter coin, i.e., $w(H) - w(T) = w(T) - w(L)$.
6. Another variation of containing one heavier and one lighter coin occurs if the counterfeit coins are unequally heavier and lighter, i.e., $w(H) - w(T) <> w(T) - w(L)$.

Thus, in solving two counterfeit coins problem an algorithm or a minimal set of algorithms have to be there that should cover all of these six above-mentioned cases separately or as a whole and thereby solving the problem in general. Previously, we have developed four such algorithms which solve the two counterfeit coins problem considering all of the six factors. We may also visualize it in the following way. Suppose we consider the weight of a standard or true coin as a positive real number (may be any value) and also the weight of a fake coin (there are two false coins) as any value other than the weight of the true coin but within a range of proportionate weight, i.e., another positive real number. The weights or real numbers of these two false coins must fall into one of the six categories, and thus, generalize it.

In moving to three counterfeit coins problem, i.e., there are now three fake coins in the search space, we at first discuss all the possible cases that may occur toward generalization and then develop an algorithm in its natural flow. Let Ci denotes the ith false coin. As there are three false coins, either heavier or lighter, C is H or L and $i = 1, 2, 3$. Again $w(Ci)$ is the weight of the ith false coin. Depending on the combination of weight of these false coins different situations may occur.

1. The three false coins present in a set of coins are equally heavier, i.e., $w(H_1) = w(H_2) = w(H_3)$.
2. All the three false coins are heavier than a true coin but they have different weights, i.e., they are unequally heavier. Thus, $w(H_1) <> w(H_2) <> w(H_3)$.
3. Two false coins out of the three are heavier and they are of the same weight and the remaining one false coin has a different weight than the other two (the weight of which is undoubtedly more than the true coin but it is either greater or less than the other two false coins). Thus, $w(H_1) = w(H_2)$ but either $w(H_3) > w(H_1)$ or $w(H_3) < w(H_1)$.
4. The three false coins present in a set of coins are equally lighter, i.e., $w(L_1) = w(L_2) = w(L_3)$.

5. All the three false coins are lighter than a true coin but they have different weights, i.e., they are unequally heavier. Thus, $w(L_1) < > w(L_2) < > w(L_3)$.

6. Two false coins out of the three are of the same weight and the remaining one false coin has a different weight than the other two (the weight of which is certainly less than a true coin but it is either greater or less than the other two false coins). Thus, $w(L_1) = w(L_2)$ but either $w(L_3) > w(L_1)$ or $w(L_3) < w(L_1)$.

7. Two false coins are equally heavier and other one is lighter such that if $\Delta w(H_i) \equiv w(H_i) - w(T)$ and $\Delta w(L_1) \equiv w(T) - w(L)$, then $\Delta w(H_i) = \Delta w(L_1)$.

8. Two false coins are equally heavier and other one is lighter such that $\Delta w(H_i) < > \Delta w(L_1)$.

9. Two false coins are unequally heavier and the third one is a lighter such that $w(H_1) < > w(H_2)$ and $\Delta w(H_1) = \Delta w(L_1)$ or $\Delta w(H_2) = \Delta w(L_1)$.

10. Two false coins are unequally heavier and the third one is a lighter such that $w(H_1) < > w(H_2)$ and $\Delta w(H_i) < > \Delta w(L_1)$.

11. Two false coins are equally lighter and the other one is heavier such that if $\Delta w(H_1) \equiv w(H) - w(T)$ and $\Delta w(L_i) \equiv w(T) - w(L)$, then $\Delta w(H_1) = \Delta w(L_i)$.

12. Two false coins are equally lighter and the other one is heavier such that $\Delta w(H_1) < > \Delta w(L_i)$.

13. Two false coins are unequally lighter and the third one is heavier such that $w(L_1) < > w(L_2)$ and $\Delta w(H_1) = \Delta w(L_1)$ or $\Delta w(H_1) = \Delta w(L_2)$.

14. Two false coins are unequally lighter and the third one is heavier such that $w(L_1) < > w(L_2)$ and $\Delta w(H_1) < > \Delta w(L_i)$.

Thus, we may conclude that a true coin as well as a fake coin may take any real number as its weight and the instance must satisfy one of the above-mentioned cases.

We can also visualize the counterfeit coin problem in a graphical point of view. A coin represents a node $(v_i, i = 1, 2, \ldots, n)$ and there is an edge between any two nodes, say v_i and v_j, if the coins representing these nodes are of equal weight and each edge weight represents that value. Hence, in the sample space all the true coins always form a complete graph whereas the false coins are disconnected from the true coins forming either isolated vertices or connected among themselves depending on their weights. Thus, this problem now reduces to graph construction problem or finding hidden graph problem. As all the coins are identical in their appearance and we do not have their weights, we may assume that all the n nodes form a complete graph, i.e., we do not have the information whether a node is true or false. At this point, the graph reconstruction problem is to find out the false edges and accordingly remove these to obtain the preferred graph (Fig. 1).

In developing an algorithm for three counterfeit coins problem, in this paper we consider two cases, i.e., there are three equally heavier or three equally lighter false coins in a set of n coins.

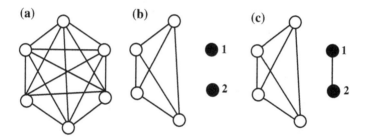

Fig. 1 **a** 6 coins or nodes with identical appearance form a complete graph. **b** Nodes 1 and 2 are false nodes with mutually different weight other than the standard (or correct) weight. **c** Nodes 1 and 2 are false having mutually equal weight

4 Algorithm Development

Through the problem specification we know that the number of false coins is three and this leads to the difficulty of checking all the possible pairs of locations among the set of n locations (for n coins). As we are considering the case where three false coins are equally heavier (or lighter), we start with four coins in the search space. Thus, among four coins three false coins may be found out as the lighter most (or heavier most) coin is the true coin according to the specification. In this case, we weigh four coins keeping two coins at each pan of the equal arm balance. The heavier pan certainly contains two of the false coins (if we consider the false coins to be heavier than the true coin) while the lighter pan contains the third heavier coin. A further weighing between the two coins on the lighter pan identifies the third heavier coin.

We may also find three false coins of this specified type among five coins. If $n = 5$, we put two coins on either pan at the beginning leaving one coin outside of the weighing. If two pans are equal in weight, both of them contain one false coin each and the third false coin must be the coin outside. Thus, depending on the results of subsequent weighing we derive the possible false coins for a set of small number of coins. But, as n increases, if the subsequent comparisons are not logically bounded, we cannot choose a specific set of coins for the next comparison. Our goal is to reduce the sample space at each level of comparison and we must take a smaller set of coins than the previous. Here, we observe an important fact about the divisibility of any integer by division of 3.

We recall the fact that any positive integer n can be classified into any of the following three cases: (i) n is divisible by 3, that we can denote as $n|3$, (ii) $n + 1$ is divisible by 3, i.e., $(n + 1)|3$, and (iii) $n-1$ is divisible by 3. Therefore, it can be easily checked to which class the given set of coins belongs to and specifically which variation of the algorithm can be applied to the provided set of coins. We take an assumption that the coins are indexed by natural numbers, i.e., 1 through n. The algorithm starts by dividing the coins into three sets $K1$, $K2$, and $K3$ such that the sets $K1$ and $K2$ contain equal number of coins. At first $K1$ and $K2$ are placed on the arms for weighing. Depending on the outcome of this weighing and the specification of the false coins, we make a conclusion to select the set that contains the false coin(s).

For the first case, each of $K1$, $K2$, and $K3$ contains $n/3$ coins. For the second case, $|K1| = |K2| = (n + 1)/3$, i.e., each of these sets have $(n + 1)/3$ coins, and $|K3| = n-2(n + 1)/3 = (n-2)/3$. So, there is a disparity of one coin between $K1$ or $K2$ and $K3$. For the third case, we calculate the number of coins in the three sets as $|K1| = |K2| = (n-1)/3 + 1 = (n + 2)/3$ and $|K3| = n-2(n + 2)/3 = (n-4)/3$. Thus, $K1$ or $K2$ contains two coins more than that of $K3$. After creating sets $K1$, $K2$, and $K3$, $K1$, and $K2$ are placed into the pans of the arm balance. Depending on the conclusion, three adaptations of the algorithm go on toward the next weighing taking different sets. At each internal node some weighing is executed considering the result of its ancestor node (whether the left pan is heavier, lighter, or of equal in weight than the right pan). Now, we like to reveal here that in the development of the algorithm, we use two functions: 1CPH(K) (1CPL(K)) and 2CPH(K) (2CPL(K)), where K stands for a set of coins.

The function 1CPH(K) (1CPL(K)) is used to find one counterfeit coin which is heavier (lighter) than a true coin in the set of K coins. We call this function only when it is sure that only one false coin is present in a set of K coins. This is essentially a binary search in context of finding the heavier (lighter) coin in the specified set and finds the false coin in O(log n) comparisons. If the cardinality of the set is even, it is divided into two subsets of equal size and these are weighed. It cannot result in equality as there is certainly one false coin. In the next iteration, we proceed with that set to which we are interested, i.e., whenever we are searching for the heavier (lighter) coin, we take only the coins of the heavier (lighter) pan. On the other hand, if the cardinality of the set is odd, we keep one coin aside and proceed in the same way as stated above with the remaining even number of coins. In this case, equality in weight between the pans may arise leading to the conclusion that the coin out of weighing is the false one. At this point, if we are sure that one particular set contains one false coin, but we are not confident of the type of the false coin, it is not a kind of binary search; hence, it does not reduce the search space into half of the previous after each weighing.

The function 2CPH(K) (2CPL(K)) finds two counterfeit coins in a set of K coins [10]. It is applied on a set of (K) coins while the subsequent results of weighing identify that set of coins to contain two equally heavier coins. This algorithm finds two false coins in O(log n) time. Here we denote the method of finding three counterfeit coins in set Ki as 3CP(Ki). Whenever this method is applied on a specific set, the cardinality of the set is checked whether $n|3$, $(n + 1)|3$, or $(n-1)|3$, where n is the cardinality of the set. Accordingly, the algorithm takes appropriate step.

In case of $n|3$, we first subdivide the coins into three sets having same number of coins as has been already discussed. After the weighing of $K1$ against $K2$, we proceed toward the next step depending on the result achieved in the first weighing. If the weight of $K1$ is greater than that of $K2$, certainly at least one false coin resides in set $K1$. But, we cannot immediately conclude the exact number of false coins in

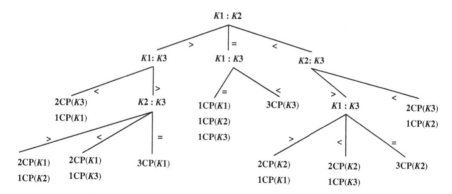

Fig. 2 Decision tree for the case $n|3$

that set as we have to consider all possible cases that may occur at this circumstance. Four possible cases are there that satisfy $w(K1) > w(K2)$.

- All the three false coins (heavier) in set $K1$.
- Two of the false coins are in set $K1$, whereas the third one is in set $K2$.
- Two of the false coins are in set $K1$, whereas the third one is in set $K3$.
- One of the false coins resides in $K1$ while other heavier coins are in $K3$.

However, here is one important thing to observe that irrespective of the case at least one false coin is there in set $K1$. Now, we weigh $K1$ against $K3$ to check whether it satisfies case four or not. If $w(K1) < w(K3)$, we conclude that case four is true and accordingly we apply 1CP($K1$) and 2CP($K3$) [7]. On the other hand, if $w(K1) < w(K3)$, we have to proceed further to identify which of the remaining cases is true. For this we compare between $w(K2)$ and $w(K3)$ and according to the result of this comparison, we reach to some conclusion for applying 2CP() and/or 1CP() on particular set of coins as depicted in Fig. 2. On the other hand, at the first level of weighing, if $w(K1) < w(K2)$, we may take the same assumption for set $K2$, as we have considered for $K1$ in case of $w(K1) > w(K2)$. The decision tree of Fig. 2 shows the subsequent comparisons. Now, we come to the third possibility that may occur at the first level of weighing, i.e., $w(K1) = w(K2)$. Two possible cases are there to satisfy this condition,

- There are equal number of heavier coins in both the sets $K1$ and $K2$. In that case, as the total number of false coins is three which is an odd number, we may conclude that $K1$ and $K2$ contain one false coin each while the third one is in $K3$.
- Two sets $K1$ and $K2$ contain only true coins having all the false coins in $K3$.

Accordingly, we weigh $K1$ against $K3$ and we apply 1CPH() on each of the sets $K1$, $K2$, and $K3$, or 3CPH($K3$), if $w(K1) = w(K3)$, or $w(K1) < w(K3)$, respectively. Thus, at the leaf nodes what we have is a specified subset having either one or two or three false coins according to which 1CPH() or 2CPH() or a further recursion of the same algorithm has to be executed on that specified set of coins. As has been

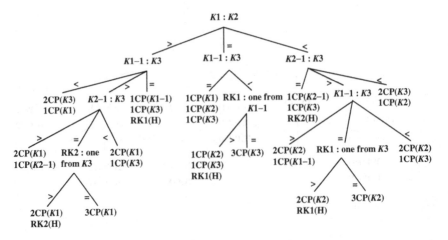

Fig. 3 Decision tree for the case $(n + 1)|3$

already discussed, each of the algorithms 1CPH() and 2CPH() is O(log n) time computable and we reach the leaves of the tree through a number of constant time comparison that reduces the time complexity of the algorithm 3CPH() into O(log n). The recursive application of 3CPH(K_i), where K_i is a reduced set from the initial set of coins, at first the cardinality is checked to find in which category it belongs to, i.e., $n|3$, or $(n + 1)|3$, or $(n-1)|3$, and accordingly the algorithm progresses.

Now, we discuss the scenario that comes when $(n + 1)|3$. In this case the initial subdivision does not result into three subsets having equal cardinality, i.e., $K1$ and $K2$ contain equal number of coins whereas $K3$ has one less coin than each of the remaining two. There is an observation that the algorithm is logically same for all the three categories of divisibility by three, i.e., it divides a set, compares between two specified sets, draws some assumptions, and upon which it reduces sample space by choosing some specific sets to be compared in the next level. The only difference between the three categories is in the cardinalities of the sets to be compared. For an example, from the decision tree of Fig. 3, after the first level of weighing we have to compare between $K1-1$ and $K3$, where $K1-1$ is the set excluding the last coin from $K1$. Again, after the second level of comparison, if w $(K1-1) > w(K3)$, set $K1-1$ certainly contains a false coin, but confusion arises when two other coins as those may reside either in $K2$ or in $K3$ or it is the remaining coin of $K1$. Thus, to verify the two sets $K2$ and $K3$ we compare $K2-1$ with $K3$. If w $(K2-1) > w(K3)$, we may conclude that at least one false coin resides in $K2-1$. Again, $K2$ cannot contain more than one false coin as $K1$ must contain one more false coin than that of $K2$ to satisfy the condition $w(K1) > w(K2)$ at the first level of weighing. So, one false coin is in $K2-1$ and the remaining two false coins are in $K1$. On the other hand, if $w(K2-1) < w(K3)$, two false coins are in $K1$ and the third heavier coin is in $K3$. If $w(K2-1) = w(K3)$, both of them cannot contain one false coin each as $K1$ contains one more false coin than $K2$. Hence, there are two possibilities.

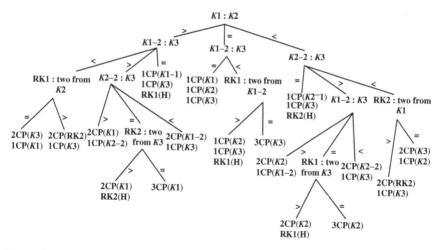

Fig. 4 Decision tree for the case $(n-1)|3$

- Both of them contain only true coins while all the false coins are in $K1$.
- Two false coins are in $K1$ whereas the remaining coin in set $K2$ is the third heavier coin (Fig. 3).

However, here we can conclude that $K3$ contains only true coins. Hence, we may believe that any coin belongs to set $K3$ as a standard coin, and we compare between the remaining coin of $K2$ and one coin from $K3$, and depending on the result we apply either 1CP(), or 2CP(), or 3CP() on the specific sets.

For the right branch of the tree of Fig. 3, the assumptions and the corresponding steps are identical taking the set $K2$ instead of set $K1$. For the equality condition, we compare $w(K1-1)$ with $w(K3)$. If $K1-1$ and $K3$ are of equal weight, we conclude that both of them contain one false coin each while the remaining coin of $K1$ is the third heavier coin. On the other hand, if $K1-1$ weighs less than $K3$, two possibilities are there; either $K3$ contains all the false coins while $K1$ and $K2$ are the true coins' sets or $K3$ contains one false coin and the remaining coin of $K1$ is one another false coin and the third one is in $K2$. In both the cases, it is convenient that $K1-1$ does not contain any false coin; hence, any of the coins belonging to this set can be considered as a standard coin. To verify which of the possibility is true, we weigh the remaining coin of $K1$ against one coin of $K1-1$ that identifies the fact.

In case of $(n-1)|3$, the algorithm flows in the same way as that of the case $(n + 1)|3$ despite of the fact that $K1$ or $K2$ contains two more coins than $K3$ (Fig. 4). After the first level of weighing, assumptions are also same as that of the previous two cases. In the second level, we compare $K1-2$ or $K2-2$ with $K3$ to have equal number of coins on both pans of the arm balance. The difference of this version from that of the previous one comes into account after the second level of weighing, following the left and right branches from the root of the tree. If the weight of $K1-2$ is less than that of $K3$, we cannot immediately conclude that $K1$ contains one false

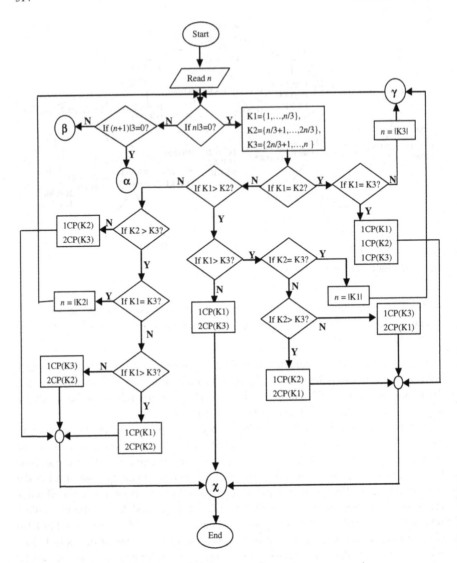

Fig. 5 Flowchart of the algorithm to find three anomalous coins among n coins

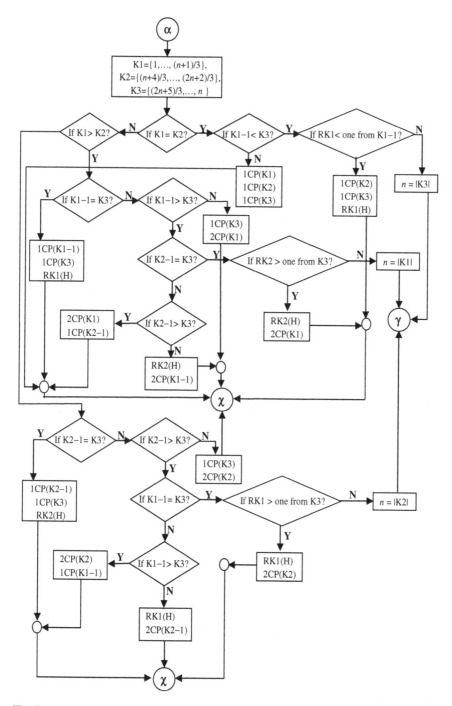

Fig. 5 (continued)

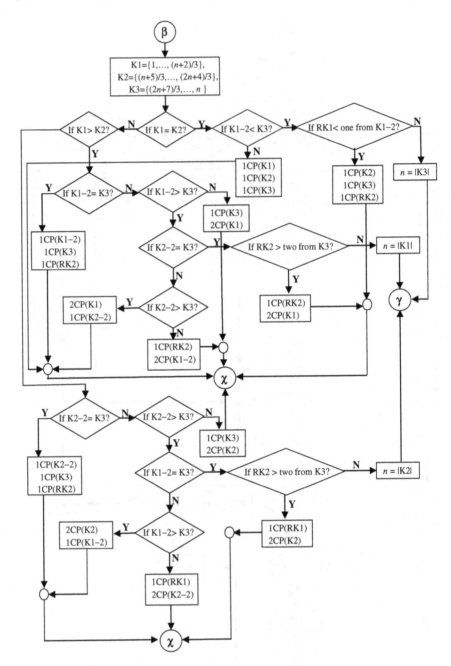

Fig. 5 (continued)

coin while $K3$ contains two false coins, as there are two remaining coins belonging to $K1$ outside the comparison. Two possibilities satisfy this condition.

- $K1$ contains one false coin while $K3$ contains two false coins.
- Remaining two coins of $K1$ are heavier and one of the coins in $K3$ is the third heavier coin.

Hence, we have to check the remaining coins with two true coins. As the coins in $K2$ have been proved as true coins, we use two of them as standard coins to perform the verification operation. Accordingly, the decision is taken to apply 1CPH() or 2CPH() on appropriate set(s) of coins (Fig. 5). In the right branch for the first level of weighing, the same process is followed depending on the result of subsequent weighing and the important thing is that the remaining coins of $K2$ are considered there instead of $K1$.

Thus, these three decision trees cover all possible sets of coins using the divisibility of three criteria. The novelty of the algorithm is not only in its generalization for all possible number of coins, but it is applicable to find out three equally lighter coins. In that case, at each node the sets of coins to be weighed are changed depending on the assumption that a pan lighter than the other certainly contains at least one of the false lighter coins.

5 Experimental Results

In this section, we discuss our algorithm in the analysis of 3CP(), 2CP(), and 1CP() with the purpose of showing the computational complexity as $O(\log n)$. In our algorithm, at the leaf nodes as we have applied 1CP(), 2CP(), or 3CP() on some precise set of coins and to attain the leaves the number of comparisons required is constant; hence, the computational complexity of the algorithm depends on the complexity of these functions. Referring the algorithm in paper [6], we choose some values of n so that it covers all the three categories for the subdivision of n and calculate the average number of comparisons requisite in the case of 2CP(). Table 1 shows the variation of required number of weighings with the number of coins under consideration. In Fig. 6, the horizontal axis denotes the total number of coins while the vertical axis refers to the average number of comparisons required to find three counterfeit coins among a set of identical looking coins. Again, 1CP() performs accurately like binary search problem, which takes $O(\log n)$ time to find one false coin among a set of n coins. To compute the average case complexity, we have to think about all the possible false coin pairs. Hence, for a given value of n, there are $^nC_3$ possible combinations.

Table 1 Average number of comparisons for some values of n

Number of coins (n)	Total number of comparisons (S)	Possible number of false coin combination C ($^nC_2$)	Average number of comparison AVG = S/C
9	171	36	4
20	1254	190	6
36	5508	630	8
54	13,365	1431	9
82	34,267	3321	10
100	54,926	4950	11
108	65,232	5778	11
144	130,842	10,296	12
198	251,883	19,503	12

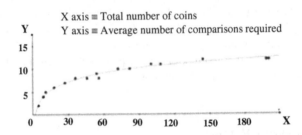

Fig. 6 Plot of average number of comparisons required against the number of coins

6 Computational Complexity

The total number of coins, i.e., n is divided into nearly three equal parts at each iteration, and the cardinality of the set on which the operations are performed, always shrinks by a factor of 3. Let us first consider the case $n|3$. As we observe at the ith level, each set contains $n/3^i$ number of coins. Now, if we achieve the set with four coins, we can solve it through two comparisons only, as we have shown earlier. So, let at the ith level of comparison the cardinality of the set decreases to four. Thus, $n/3^i = 4$, i.e., $3^i = n/4$. Therefore, $i = \log_3 (n/4)$. Moreover, if 3CPH(K_i) is applied at each iteration before getting the set with four coins, we can observe in all the decision trees in Figs. 1, 2, and 3, that to reach a leaf where 3CPH() is applied we have to traverse at most three internal nodes, i.e., at most three comparisons are required at each iteration. Hence, a total of $3 \times i$ comparisons are required resulting in $3 \times i+2$ comparisons in total, which is O(log n). As we have already explained, that to find a known type of false coin (either heavier or lighter) always takes O(log n) time. On the other hand, if 2CPH(K_i) is applied at a leaf node, it also takes time O (log n) [11]. If 1CPH(K_i) or 2CPH(K_i) is performed at j-th level of comparison it is sure that before that iteration 3CPH() is executed for ($j-1$) times. We know that 1CPH() requires at most $\lceil \log_2 n \rceil$ comparisons and at the j-th level it is to be applied on $n/3^j$ number of coins. Thus, it would take a total number of $2|j| + 2\log_3(n/3^j)$ comparisons. Hence, in the worst case it would take $O(2|j| + 2\log_3(n/3^j)) + O(2 \times \log_3(n/5) + 4)$, i.e., $O(\log n)$ comparisons altogether as j is $O(\log n)$.

7 Application

Counterfeit coin(s) problem has vast applications in different fields; some of which are discussed here. In hidden graph learning problem, where we have the information of the number of vertices as well as the number of edges but we do not know the actual edges (i.e., the vertex pairs) and we are to construct the graph through minimum number of query [10]. Again, graph finding or construction problem is the key issue in the domain of Bioinformatics and DNA sequencing.

Furthermore, coin weighing problem is exhaustively studied in anomaly detection problems in quantum information processing [11], compressed sensing, and multi-access adder channel [12]. As counterfeit coin(s) problem belongs to combinatorial group testing problem, it can be mapped into utilization in medical field like finding of any odd spike in MRI scan, or in technical field to find any set of damaged pixels in a digital image, locating electrical shorts in electrical circuits [12], etc.

8 Conclusion

The issue of counterfeits violates rational property right and also causing harm to both producer and purchaser. In this paper, we have developed an algorithm to recognize three anomalous coins among a set of n coins which are identical in exterior. In this case, we have assumed that all the false coins are equally heavier (or lighter) than a true coin. Two versions of the problem, i.e., all the counterfeit coins are equally heavier or equally lighter, can be solved using this algorithm with the time complexity $O(\log n)$. The most significant fact is that the decision tree structure can be used to solve such problems of large size, by eliminating a part of the solution domain after each step of decision making. Especially, as our algorithm works for any value of n, it does not make an issue if the value of n is not known a priori.

References

1. Ghosh, J., Senmajumdar, P., Maitra, S., Dhal, D., Pal, R.K.: A generalized algorithm for solving n coins problem. In: Proceedings of the 2011 IEEE International Conference on Computer Science and Automation Engineering (CSAE 2011), vol. 2, pp. 411–415. Shanghai (2011)
2. Ghosh, J., Senmajumdar, P., Maitra, S., Dhal, D., Pal, R.K.: Yet another algorithm for solving n coins problem. Assam Univ. J. Sci. Technol.: Phys. Sci. Technol. 8(II), 118–125 (2011). ISSN: 0975-2773

3. Tošić, R.: Two counterfeit coins. Discrete Math. **46**, 295–298 (1983) (North-Holland)
4. Manvel, B.: Counterfeit coin problems, mathematics magazine, mathematical association of America **50**(2), 90–92 (1977)
5. Bellman, R., Gluss, B.: On various versions of the defective coin problem. Inf. Control **4**(2–3), 118–131 (1961)
6. Ghosh, J., Dey, L., Nandy, A., Chakraborty, A., Datta, P., Pal, R.K., Samanta, R.K.: An advanced approach to solve two counterfeit coins problem. Proc. Ann. Pure Appl. Math. **7**(1), 77–82 (2014). ISSN:2279-087X (P), 2279-0888 (online)
7. Ghosh, J., Datta, P., Chakraborty, A., Nandy, A., Dey, L., Pal, R.K., Samanta, R.K.: An endeavour to find two unequal false coins. In: Proceedings of the 8th International Conference on Electrical and Computer Engineering (ICECE 2014), pp. 333–336. Dhaka (2014)
8. Ghosh, J., Chakraborty, A., Datta, P., Dey, L., Nandy, A., Pal, R.K., Samanta, R.K.: The first algorithm for solving two coins counterfeiting with $\omega(\Delta H) = \omega(\Delta L)$. In: Proceedings of the 8th International Conference on Electrical and Computer Engineering (ICECE 2014), pp. 337–340. Dhaka (2014)
9. Ghosh, J., Nandy, A., Dey, L., Datta, P., Chakraborty, A., Pal, R.K., Samanta, R.K.: An algorithm for identifying two unequal heavier / lighter coins out of n given coins. In: Proceedings of the 3rd International Conference on Computer, Communication, Control and Information Technology (C3IT2015), pp. 1–6. Academy of Technology, West Bengal (2015)
10. Kim, J.H.: Finding Weighted Graphs by Combinatorial Search (2012). arXiv:1201.3793v1 [math.CO]
11. Lim, E.W.C.: On Anomaly Identification and the Counterfeit Coin Problem (2009) arXiv:0905.0085
12. Du, D.Z., Hwang, F.K.: Combinatorial Group Testing and its Applications Series on Applied Mathematics, vol. 3. World Scientific Publishing Co. Pvt. Ltd., Singapore (1993)

Author Index

© Springer India 2016
R. Chaki et al. (eds.), *Advanced Computing and Systems for Security*,
Advances in Intelligent Systems and Computing 396,
DOI 10.1007/978-81-322-2653-6

Printed in the United States
By Bookmasters